IMPROVING PROFITABILITY THROUGH GREEN MANUFACTURING

IMPROVING PROFITABILITY THROUGH GREEN MANUFACTURING
Creating a Profitable and Environmentally Compliant Manufacturing Facility

DAVID R. HILLIS
J. BARRY DUVALL

Greenville, North Carolina

A JOHN WILEY & SONS, INC., PUBLICATION

Published by John Wiley & Sons, Inc., Hoboken, New Jersey
Published simultaneously in Canada

For general information on our other products and services or for technical support, please contact our Customer Care Department within the United States at (800) 762-2974, outside the United States at (317) 572-3993 or fax (317) 572-4002.

Wiley also publishes its books in a variety of electronic formats. Some content that appears in print may not be available in electronic formats. For more information about Wiley products, visit our web site at www.wiley.com.

Library of Congress Cataloging-in-Publication Data:
Hillis, David R.
  Improving profitability through green manufacturing : creating a profitable and environmentally compliant manufacturing facility / David R. Hillis, J. Barry DuVall.
    p. cm.
  Includes bibliographical references and index.
  ISBN 978-1-118-11125-3 (cloth)
  1. Manufacturing processes–Cost control. 2. Manufacturing processes–Environmental aspects. 3. Green products–Cost effectiveness. 4. Sustainable enginering. I. DuVall, John Barry, 1942– II. Title.
  TS183.H55 2013
  628–dc23
                                    2012009796

Printed in the United States of America

10 9 8 7 6 5 4 3 2

CONTENTS

PREFACE

Let us tell you our story—we were on the road, driving over to meet with people at a manufacturing plant in western North Carolina. We were early for our meeting so we thought we would stop at a fast food restaurant for coffee and a chance to review our meeting agenda. We carried our coffees back to the table, opened a file folder, and booted-up a laptop. A minute later we were surprised to find that we didn't have enough table space. Our two cups of coffee and the packaging debris of cup lids, cream containers, sugar packets, napkins, stirrers, and a tray left just enough room for one laptop or one open file folder. So we made a quick trip to the waste container to dump all this trash and leave the tray. We realized that even though all this trash was annoying it had served a purpose—it had utility. We finished our coffees and left for the meeting.

When we arrived our host asked if we cared for a coffee. We declined with a laugh and recounted our experience with "coffee waste." They wanted to know if the restaurant "recycled?" It did not but we pointed out that that was not the issue. So we mentioned the "Three Rs." It is a path for waste reduction promoted by Singapore's National Environment Agency and other agencies around the world. The "R"s stand for "Recycle, Reuse, and Reduce" packaging and convenience materials. The goal for implementing the Three Rs is to minimize the amount of solid waste that is generated daily. For us, the reality of this concept struck home on another business trip to southern California.

On that trip we stayed at a chain hotel that provides a complimentary breakfast for its guests. There were two of us at the table and by the time we finished eating the entire table was covered in packaging litter. There were single-use plastic spoons, knives, plastic foam cereal bowls, paper coffee cups, plastic juice cups, paper milk cartons, plastic fruit containers, paper sugar envelopes, napkins, and several cellophane wrappers. We picked up all this litter and placed it into the appropriate recycling containers: paper, plastic, and food refuse. We had accomplished the first R—Recycling.

When our morning meetings were finished we stopped at a corner restaurant for lunch. We had soup, salad, and coffee. This food all arrived on sturdy china plates with stainless steel flatware. There were paper napkins, but that was the only single-use item. All the other items were in the second category—Reuse.

Finally, at the end of the workday our host suggested we have dinner at an Ethiopian restaurant. There were four of us at the table and the food was brought out on a large platter. There were some bowls with sauces but little else. There were no eating utensils. We discovered that we were to use the flat bread that was served to scoop up the food. By the time we finished eating there was one large empty tray in the center of the table along with a few bowls. The restaurant had provided cloth napkins, which would be laundered and reused. The dinner was an example of the third category—Reduce.

These stories help make the point that a strategy for waste reduction should aim at moving up the hierarchy away from single-use items to a system that reduces the packaging and convenience items. Recycling therefore is not an endpoint but a starting point for the Three Rs of waste reduction. So, you may ask how do you move up this hierarchy? A partial answer is waste reduction begins with the design of the product. Recycling is accepting the current design and then trying to make the best use of the waste that is generated by that design. In our southern California trip we started the day with a "serving design" for breakfast that was predicated on recycling. By evening of that day the serving design for dinner was based on the third R, reduce. That design had less waste. So think of the Three Rs as a systems approach to waste reduction. The approach presented in this book is also a systems approach, but it is applied to manufacturing. Both approaches provide a strategy for analysis, decision making, change, and improvement. They also provide opportunities! The beauty of the systems approach is that it can be used to analyze complex things and make them simple. Albert Einstein said it well: "make everything as simple as possible, but not simpler."

In our discussions with corporate executives and environmental groups, and working with practicing professionals on the plant floor, we came to the realization that success in manufacturing is not based on magic or "green technology." We learned that traditional manufacturing companies can be environmentally responsible and profitable through improved decision making.

In this book we provide a model for improvement that you can modify and apply in your own way, in your own environment. We have provided examples of this systems approach along with supporting methods and techniques that are being used by a variety of manufacturing companies to be environmentally responsible and profitable. Now it is up to you! We hope this will be helpful to you in your company and industry. Please let us know how things go and feel free to contact us if we can help you in the future.

DAVID R. HILLIS
J. BARRY DUVALL

Greenville, North Carolina
hillisd@ecu.edu
duvallj@ecu.edu

ACKNOWLEDGMENTS

The authors thank our wives, Carol and Jean, for their encouragement and belief in the importance of this project, and the following individuals, companies, and organizations for their assistance and contributions.

Members of the North Carolina Department of Environment and Natural Resources, Division of Pollution Prevention and Environmental Assistance, particularly:

Ms. Julie Woosley, Section Chief

Ms. Angela Barger, Environmental Stewardship Initiative Lead Coach

Mr. Ron Pridgeon, Environmental Engineer

Singapore's National Environment Agency

Ms. Hong Yang, Manager Waste Minimization & Recycling, Waste and Resource Management Department

Engineered Sintered Components, Troutman, NC

Ms. Jan Comer, Vice President Human Resources

Mr. Marty Todd, Vice President Operations

Mr. Stephen Jenkins, EHS/Training Manager

Corning Cable Systems, Hickory Manufacturing and Technology Center, Hickory, NC

Mr. Steve Street, Senior Environmental, Health, and Safety Coordinator

East Carolina University, College of Technology and Computer Science, Department of Technology Systems, Greenville, NC

Thanks to our students for the opportunity to work with and learn from you, and for testing PCPCs in your own work environments:

Brian Miller, Carver Machine Works, Washington, NC

Mr. Monty Hilburn, Hamilton Sundstrand Corporation, San Diego, CA

Armen Ilikchyan, Cooper Standard Automotive Corporation, Bowling Green, OH

Gregg Phipps, Gasdorf Tool and Machine Corporation, Lima, OH

Ms. Rebecca Farmer, U.S. Coast Guard Aviation Logistics Center, Elizabeth City, NC

CHAPTER 1

MANUFACTURING

INTRODUCTION

It frequently surprises people when they learn that the world's leading manufacturing country is the United States of America. Why this may be so astonishing is the prevalence of "Made in China" labels found on so many consumer products, particularly clothing and electronics. In 2007, prior to the recession in the latter part of that decade, the value of goods produced by the United States reached over $1.8 trillion. (see http://unstats.un.org/unsd/snaama/cList.asp)—and, even more surprising, the amount produced in 2007 was nearly twice the value made two decades earlier. Today the United States is still a major producer, generating much of its prosperity from manufacturing. Nevertheless, there is no doubt that a large portion of our products come from overseas.

Part of the reason the United States continues to lead in the production of goods is the manufacturing methods or procedures

Improving Profitability Through Green Manufacturing: Creating a Profitable and Environmentally Compliant Manufacturing Facility, First Edition.
David R. Hillis and J. Barry DuVall.

that were developed during the twentieth century. These methods enabled companies to produce large amounts of affordable goods profitably. During the latter half of that century other nations adopted these methods and even made substantial improvements. Now many believe that manufacturing in the United States is too costly both in dollars and harm to the environment. This is not true. There are ways to make manufacturing sustainable and profitable while meeting environmental obligations and requirements.

MANUFACTURING SEQUENCE

To understand how this can be done let's begin by examining the *manufacturing sequence*. The production of a product begins after a raw material has been transformed into a manufacturing "stock." Think of "pig iron" as a raw material and 16-gauge cold-rolled steel as a manufacturing stock. Yes, an argument can be made that pig iron is a manufacturing stock after iron ore has gone through a smelting process. Regardless of where the starting point occurs there is a specific series of steps that occur in the manufacture of a product and its sale to a customer. Figure 1.1 illustrates these steps.

A simple example of this sequence is the manufacture of a molded plastic bowl that is actually a component that will be assembled with other parts to create a more complex product—an inexpensive food processor. The bowl is produced by a molding process using a stock of plastic pellets. The pellet stock is polystyrene, which is produced from an aromatic polymer that comes from a liquid hydrocarbon manufactured from the *raw material*, petroleum. The food processor is next distributed to a customer. After years of use the bowl cracks and the owner finds that it has a recycle number "6" discretely molded on the bottom of the bowl. The owner of the bowl deposits it in a recycling bin that ultimately allows it to be *recycled* into another stock. The manufacturing sequence in this instance is a closed loop, illustrating one of the several definitions for a *product life cycle*.

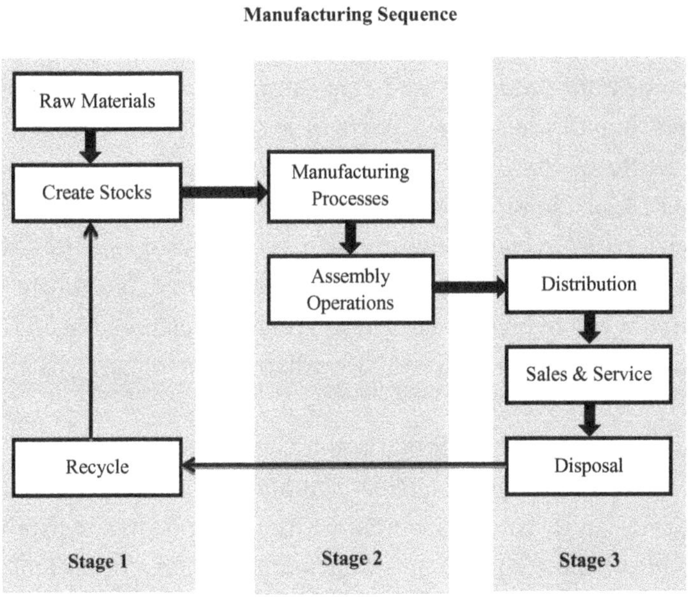

Manufacturing Sequence

Figure 1.1. The general sequence of manufacturing.

PRODUCT LIFE CYCLES—THERE'S MORE THAN ONE

This concept of a product's life cycle based on the manufacturing sequence provides a useful perspective for developing a competitive and compliant facility. However, the term *product life cycle* is also used to name several other concepts. Probably the most well-known use refers to a marketing-oriented definition of the phases or stages a product passes through over its lifetime. Marketing people generally list five phases, beginning with "product development." The next phase is the product's "introduction into the marketplace," followed by a "sales growth" phase. The last two phases are "product maturity" and finally the product's "decline" in the marketplace. In this instance the life cycle traces the life span in terms of the product's sales volume in the marketplace.

A third form of analysis that shares the title "product life cycle" includes the term "management": *product lifecycle management*

(PLM), which involves managing the information acquired over a product's life so that a company understands how its products are designed, built, and serviced. The emphasis is primarily on the engineering and business aspects of producing the product.

The title of the fourth application, *product life cycle management* (PLCM), sounds identical to the previous one. The difference of course is *lifecycle* is now two words instead of one. PLCM has to do with the strategies a business uses to manage the life of a product in the marketplace. These strategies change based on the product's "marketing phase." Recall the five phases mentioned earlier.

There may well be other product life cycle methods or techniques in use. However, this sampling illustrates their basic objective—to enable a business to understand how a product is doing in the marketplace and what improvements or actions need to be taken to increase sales, performance, and/or safety. These techniques are used primarily for increasing a company's profitability. Our objective, however, is to improve both the company's profitably and its environmental performance.

To do this we'll go back to the general sequence of manufacturing. Recall the example involving the plastic bowl? The bowl started out as a raw material and moved through the manufacturing sequence until it was purchased and placed into use. When it cracked it was recycled. This sequence can be used as the basis for an analysis that examines how manufacturing impacts the environment: life cycle analysis (LCA).

LIFE CYCLE ANALYSIS

The origins of *life cycle analysis* probably came from the environmental impact studies and energy audits that were carried out in the late 1960s and early 1970s. These studies attempted to assess the resource costs and environmental implications of the industrial practices going on in the world at that time. Paper manufacturing, as well as its associated recycling processes, was one of several activities that received a great deal of attention in these early studies. The methods these studies used were unique at the time because they followed the entire sequence of business. As with the manufacturing sequence, these studies started with turning raw materials into usable stocks for production and followed the sequence through distribution, the customer's use, and finally the product's disposal or recycling. The analysis attempts to identify the environmental costs associated with a product by examining the all the resources and materials used along with the wastes released to the environment over a product's lifetime.

These studies have evolved into a defined protocol. The LCA has become a popular technique in building and construction projects. In fact its popularity has reached a level that there are several software products available to assist in the analysis. An example is the "Building Life-Cycle Cost" (BLCC) program developed by the National Institute of Standards and Technology (NIST). The U.S. Department of Energy's Federal Energy Management Program says that the BLCC enables architects and builders to evaluate alternatives to find the most cost-effective building designs in terms of energy use over the life of the project.

Along with LCA and BLCC there are a variety of other terms being used to describe this technique. The most familiar term is probably LCA, but there are others now in use such as *life cycle-inventory* (LCI) and *life cycle assessment* (also abbreviated LCA). Also, if you do an Internet search on LCA you will also find more terms such as cradle-to-grave analysis, eco-balancing, and material flow analysis. Regardless of the name, the primary aim of life cycle analysis is to identify the environmental impact of the materials and resources used in the manufacture and use of a product.

To be of value the analysis needs to identify and quantify the source and amount of waste generated over the entire manufacturing sequence. This is similar to a procedure that financial managers call *sources and uses*. Large publicly traded companies will include a "sources and uses of funds" statement in their annual reports. The resource in this case is money—where it is obtained, its source, and how it is used to carry out the activities of the business. Individuals and institutions that are contemplating lending money to a startup company look for a sources-and-uses worksheet because it is an excellent summary of the "startup's" financial plan. In a similar manner an LCA can be viewed as a sources-and-uses statement.

Most LCAs include a comprehensive listing of the inputs, the resources. The output defines how effective the facility is in converting these resources into products while minimizing waste. Inputs include all raw materials, stocks, and resources that are used for the creation of the product. Resources include energy demands (electricity, gas, oil, coal, etc.) and water. In some special instances land use might be included. While land is not considered a consumable in the creation of a stock or product, there could be a circumstance that would make the land unusable for a period of time. An example is strip mining.

Figure 1.2 shows an example of an LCA format. This format includes the most common steps in the manufacturing sequence plus extraction of raw materials and repair or service. The outputs of course include the product as well as everything else, which is defined as waste. The major waste categories are water and water effluents; airborne emissions; solid waste; and recyclables. An item that is often overlooked in the analysis of manufacturing waste is packaging. This is not the case in LCA. A major source of waste in the distribution step of an LCA is "single-use" packaging.

The LCA, like the manufacturing sequence, addresses material in the first two steps. On the left side of Figure 1.1 these steps are identified as Stage 1. In the first two steps of an LCA (extraction of raw materials; creation of the stock) the industry carries the name of the material being converted to a stock. As an example, when someone says "the steel industry" what comes to mind? In most cases an image of a steel mill will pop up in our mind's eye.

Manufacturing Sequence

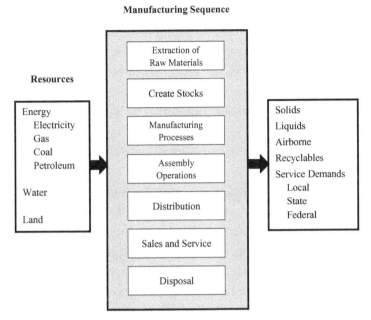

Figure 1.2. A life cycle analysis model.

However, when it becomes a coil of cold-rolled steel it is a stock that will be used to manufacture a product. So, beginning with the third step (Stage 2 in Fig. 1.1) of the LCA, the industry name changes from the material name to the product name. Steel would be replaced by a product name such as auto or appliance.

It is apparent that completing an LCA on just one group of materials or a single industry such as the appliance industry is a major undertaking involving hundreds of companies. However, it is the approach that the analysis uses that is valuable.

Measuring and quantifying the costs of all the materials and resources required to create a product is a basic part of manufacturing. Cost accountants have been allocating direct and indirect costs to specific products and work centers for more than a century. An example of a direct cost is the amount of a stock used to create a product. These direct costs and their proper allocation to a product are relatively easy to calculate. Indirect costs are more difficult to assign. They are expenditures that are not apparent by examining the bill for materials or the list of operations used to make the product. The classic example of an indirect cost is the person who sweeps the aisles when a shift is over. Accountants often handle these costs by allocating them as a percentage of floor space used in the plant to produce a specific product or by using some other proportionality. The task, which is also the problem, is developing a method that will account for all the stocks and resources and then accurately apportion them to the product.

A further complication when using LCA is its "comprehensive" approach. The point-of-view taken by an LCA is excellent. It is the environmental version of the sources-and-uses worksheet but it is applied on an industry scale—much too general for a company involved in just one step of the manufacturing sequence. The general approach of the LCA, however, would be useful for building a new plant. It is not difficult to list the activities at each of the seven steps of a manufacturing sequence for constructing a manufacturing plant. You can list the stocks and processes used to construct the building and the assembly operations to put in the electrical distribution system, the HVAC, the plumbing, and so on that are needed to complete the facility. Servicing the building and its final disposal can also be handled effectively. Therefore the LCA is an excellent technique for assisting management in costing and designing an environmentally effective manufacturing facility.

However the LCA doesn't adapt very well for a company that makes, for example, impeller blades for diesel fuel pumps and dishwashers. The overall approach of the LCA doesn't provide a means to identify or quantify the value of the alternatives available for improving profits and becoming environmentally compliant. The question then becomes how can the LCA concept be used? Chapter 2 introduces an alternative approach that carries with it the underlying notion of an LCA. It is founded on a detailed examination of the waste and resources required to process the materials to manufacture a product.

POTENTIAL FOR WASTE AND VALUE ADDED IN MANUFACTURING

Each of the seven major activities in the manufacturing sequence offers manufacturers opportunities for creating value and waste. Table 1.1 lists these opportunities along with their potential for creating waste. This potential will vary for each of the seven steps. For instance, an assembly operation may generate some waste but generally the environmental impact will be minimal. However, in some of the other steps the waste and environmental costs can be quite high. As an example for the extraction of raw materials a large part of all waste will be environmental costs.

The table's third column lists the *value-added potential* for each step in the manufacturing sequence. As is the case with waste, the potential to add value varies significantly depending on the step and certainly on the product being made. You'll notice that assembly operations have a low to moderate potential for waste and a moderate potential for adding value. Balancing the potential for waste against the potential for adding value has been a manufacturing tactic for years. Changes in technology and proprietary knowledge can also reorder the balance between value added and waste generated for a particular step.

TABLE 1.1. Potential for Creating Waste Compared with the Value-Added Potential for Each Step in the Generalized Manufacturing Sequence

Manufacturing Sequence	Potential for Creating Waste	Potential for Adding Value
Extraction of raw materials	High	Moderate
Create stocks	Moderate to high	Moderate
Manufacturing processes	Moderate to High	High
Assembly operations	Moderate	Low to moderate
Distribution	Low	Low
Sales and service	Low	Moderate
Disposal	High	Low

A company that limits itself to performing just one step in the sequence would in theory be simplifying its business by focusing on just that function. However, this limits the company to the amount of value added in that step. Alternatively a manufacturer could try to do all seven steps and earn all of the value-added potential from the raw material to the sale and disposal of the product. Of course all the potential for waste would be present too. Also, the company would have to develop the skill and expertise for all aspects of the manufacturing sequence. There is a company that became the classic example of this approach.

At the beginning of the twentieth century Ford Motor Company had success in producing a rugged and durable automobile. The car design was good but there were other autos being manufactured at the time that were just as good. Henry Ford, however, wanted to make large numbers of cars that were affordable. With this in mind he toured plants in other industries to understand how they made their product. It has been mentioned that he came up with the idea of a continuously moving production line after he had visited a meat packing plant. True or not, he eventually concluded that effective large-volume manufacturing has four principles:

- The product uses interchangeable parts; no custom fitting or modifications should be required.
- The product moves to each workstation at a predetermined rate; this was the introduction of continuous flow manufacturing.
- The work to manufacture the product should be broken into a sequence of simple easy-to-learn tasks.
- Reducing or eliminating waste of all kinds is an ongoing effort.

It took Ford five years to put these four principles into operation; that was in 1913 at his plant in Highland Park, Michigan. These changes created the first moving assembly line ever put into service for large-scale manufacturing. Very quickly the assembly line became the icon for Ford's system of production.

A year later the continuously moving assembly line had significantly increased production and labor productivity. However, Ford's monthly turnover of labor had reached 40 to 60 percent. The company realized that this was due largely to the tedium of assembly-line work and the frequent increases in the production quotas that were placed on the workers. Ford solved this turnover problem by paying his workers $5 a day when other manufacturers were paying about $2.50 per day. The increase in labor costs were offset by an increase in output (productivity) due to a more stable workforce. The improved productivity also provided a substantial increase in the company's profits. At the same time the company's profits were increasing, the price of the Model T continued to drop. The result was an increase in demand for the Model T. Before Ford stopped making the Model T in 1927 over 15 million of these cars had been sold.

VERTICALLY VERSUS HORIZONTALLY INTEGRATED MANUFACTURING

Certainly the manufacturing principles that Henry Ford and his team developed were important. But it shouldn't be overlooked that the company also had an intense commitment to lowering costs and capturing all the value-added opportunities available in making and selling automobiles. His company embodied most of the steps in the sequence of manufacturing, starting with mining the iron ore to create the steel stock that went into their cars. Ford's enormous industrial facility on the Rouge River in Dearborn, Michigan, took the iron ore off ships and just days later the ore was steel and iron in finished automobiles on the way to car dealers. The Ford Motor Company was an excellent example of a *vertically integrated manufacturer* and for its time a lean manufacturer. Recall Ford's fourth principle.

At the same time Ford was developing his system of manufacturing automobiles, most of the other car builders remained specialists, concentrating on some processing but primarily on assembly. These companies limited themselves to just one or two steps in the sequence of manufacturing. Therefore they could be described as being more nearly *horizontally integrated*. Usually a horizontally integrated manufacturer makes more than one product. The company can expand its production or sales by offering a wider range of products or models. If the company elects to purchase parts and limit its manufacturing operations to just assembly, then it becomes a *limited horizontally integrated* manufacturing company.

There is an interesting observation that can be made. Ford's innovations were primarily in the way cars were made—Ford developed manufacturing technology, not product technology. The change in product technology was modest during the period the Model T was in production, which started in October 1908 and continued until 1927 when its replacement, the Model A, was introduced. During this period Ford was able to master the complexity of the entire manufacturing sequence and capture most of the value-added opportunities.

So what is the downside for vertically integrated manufacturers? These companies are much more vulnerable when product technology is changing quickly. A horizontally integrated manufacturer can usually adopt new technology more quickly than a vertically integrated company. In part that's because the horizontally integrated manufacturer has a narrow focus, reducing the knowledge and skills that must be acquired. Similarly the amount of investment needed for new equipment and tooling is also less. The magnitude of change for the vertically integrated company can be enormous. They tend to "hang on" to processes and methods that are not competitive, thereby forcing them into a period of being unprofitable and relying on costly "stopgaps" to be environmentally compliant.

The personal computer (PC) industry provided a good example of the impact that fast-changing product technology can have on manufacturing. In the late 1970s several of the desktop PCs relied on an agglomeration of parts that included portable cassette-tape players for data storage. Assembly was very basic and similar to the methods used to produce a television set. In fact many of the popular desktop computers at the time were actually electronic kits that were assembled by hobbyists and technicians. The key component was the microprocessor, which was a single chip that replaced all the circuitry that formerly occupied large cabinets in mainframe computers.

By the mid-1980s there appeared to be an opportunity for a sophisticated vertically integrated computer manufacturer to enter the PC market. The PC market fueled by the popularity of word processing and spreadsheet programs was definitely in the sales growth phase. One such company that recognized this opportunity was IBM, which at the time spanned at least three of the seven steps in the manufacturing sequence of a PC.

However, the PC was being produced during a period of rapidly changing product technology. Most of the components used in a PC were produced by companies that specialized in just one step of the manufacturing sequence. When innovations in data storage occurred such as in "disk drives," the producers of PCs quickly adopted the new style "floppy disk" into their product. By the late 1980s the technical product life of PCs was measured in months, which meant the distribution step became critical in the sequence of manufacturing. Dell computers exploited this by selling directly to the PC user. During the early 1990s horizontally integrated companies became dominant as producers of PCs.

WASTE AND ITS UNEXPECTED SOURCES

Regardless of whether a manufacturer is involved in one or all of the steps in manufacturing, the fundamental strategy for a company should be to maximize value added by minimizing waste. Obviously during periods of rapid change in product technology a company might be wise to limit its manufacturing involvement to two steps, assembly and distribution. Products such as the tablet computer and the smart phone provide examples of how this strategy can work, especially while these products are in a growth phase in sales. However, product technology is normally an evolutionary process. It can creep up on companies particularly when their products are in a mature phase in sales. Too often companies feel that the most effective way to differentiate their product and maintain sales levels is through price, specifically price reduction. After the price is reduced the company then looks for ways to reduce its manufacturing costs so that it can remain or once again become profitable.

Too often organizations try to reduce cost by taking away value from the product. That is wrong. Reducing value by eliminating features, service life, or functionality is placing the burden of cost reduction on the customer when it should be the organization's responsibility. The focus of cost reduction must be on the elimination of waste. In later chapters specific methods for identifying opportunities for waste reduction are introduced. There are also some case studies to illustrate how some of these waste reduction methods are used by companies. However, before moving on to these topics the source and types of waste need to be defined.

The First Source of Waste

The first major source of waste originates in the way the company makes its products. Much waste is due to the product design and the manufacturing processes used in the plant. Certainly the type of materials a company uses, which is a function of the product design, will dictate the plant design and the processes. Each material and the associated manufacturing processes have their own set of waste parameters that defines the facility. But some of the waste occurs due to the organization of the facility and the operation norms that have been established. These are the fixed sources of waste that are seldom challenged. The following are examples of items that would contribute to this source of waste:

- *Resources.* The fuels needed to operate the processes, machines, and equipment used in the manufacturing sequence.

This would also include plant and office heating, lighting, and air conditioning.

- *Water.* This also includes the associated costs of sanitary and storm sewer services.
- *Supplies.* The secondary materials that are required to complete a manufacturing operation or process but do not become part of the product. An example might be cutting tool coolant/lubricant, towels, cleaners, copy paper, etcetera. These are waste materials that are "accepted" as being part of the manufacturing operation or process.
- *Wages Paid.* Payment to individuals and to contractors or suppliers who do not add value to the product. This is one of the most difficult categories to control.

It is unusual for manufacturers to think of the items in this list as waste. Often companies "see" these as givens, inherent and necessary to conducting their manufacturing operations and certainly not waste. Can this source of waste be eliminated? No, but there is waste that can be eliminated. Therefore the first step is to recognize that there is waste in this listing and then take the second step—determine the magnitude of the waste in these four categories.

Measuring resource costs and water use for specific manufacturing operations can be done with the help of technology in the form of *smart meters.* Devices of this type can provide the "where, when, and how much" for resources being consumed. These meters are often referred to as time-of-use meters. Currently the most widely used smart meter is for monitoring electrical consumption; however, similar devices exist for measuring natural gas and water use. In large installations these devices can be networked to provide for real time monitoring and control of consumption. There are several software programs available that can make it relatively easy to complete a detailed analysis for the dollar cost of resources per unit of production.

Supplies frequently receive a great deal of attention during recessions and business downturns. Often more attention than they warrant. This category needs to be judged using the *Pareto principle*. You may recall that this principle is based on the work of Italian economist Vilfredo Pareto, who observed at the beginning of the twentieth century that 20 percent of the people owned 80 percent of the wealth in Italy. He and others found that this distribution disparity occurs frequently. As an example the "eighty–twenty rule" as it is called may apply to customers: 80 percent of a company's sales go to 20 percent of its customers. So to put this principle into practice one needs to concentrate on the significant few—examine the major costs first and ignore the trivial many. The costs of supplies are often among the trivial many. However, before supplies can be dismissed they need to be assessed to make sure they are not one of the significant waste streams that demand immediate attention.

Identifying wages paid to people who do not add value to the product is a *can of worms*. The definition of the term "can of worms" describes the circumstances beautifully. It is a complex troublesome situation, a mess of entanglements arising when decisions or actions produce subsequent problems. This is why this source of waste is seldom addressed. Being identified as not adding value to your company or branded as a "source of waste" is definitely going to create a troublesome situation. Consequently this item has to be dealt with carefully.

To start we need to examine the definition of a value-added task. The strictest connotation states that a value-added task is an activity that changes or transforms a product or moves the product closer to the consumer. A more general definition recognizes indirect activities or tasks that add value to a product. Examples of these indirect tasks are found in purchasing, production planning, maintenance, billing, payroll preparation, and so forth. These are all support activities that don't touch the product. However, these activities must be carried out effectively for a manufacturing facility to perform efficiently. So the question arises: what is a non-value-added job? Some examples are "inspectors" trying to find defective parts or products, material handlers moving parts and products that can't be shipped, and managers meeting with employees trying to settle grievances. It can be argued that these three examples of non-value-added tasks just mentioned are necessary to be an effective manufacturer; however, they are definitely symptoms of a second source of waste.

The Second Source of Waste

The second source of waste comes directly from manufacturing operations. This source has received the most attention since factories were established to produce products. Over the past few decades the people involved in controlling this source of waste now include manufacturing operations and engineering; human resources; and training and development.

There are a lot of names and acronyms for programs, techniques, and tactics used to reduce waste in manufacturing. The concept of *lean manufacturing* has become the most comprehensive approach for waste reduction now being employed. In general a lean manufacturing program works to reduce eight types of waste:

A starting point for reducing waste is to conduct a "check-up" not unlike the way a physician conducts an annual physical. There are tests to be conducted and information to be gathered along with a physical examination. The manufacturing facility has to be examined in a similar manner. A helpful approach is to use a checklist for developing a diagnosis—an assessment of the problems from the eight sources of waste:

1. *Waste from Overproduction.* More products are produced than required by the customer. Often this is done to reduce idle time or to make up for anticipated defects or product losses; excess production also may result from incentive systems.

2. *Waste from Transportation.* There may be excessive movement of the product or its components during the production process. In practice this is moving work-in-process into and

out of temporary storage or moving work because of poor facility arrangement.

3. *Waste of Motion.* Waste of motion occurs when the operator has to look for tools or information, to make adjustments or repairs, to free jams, or to fill out incentive tickets or routing sheets.

4. *Waiting.* Time may be wasted, for example, waiting for setups to be completed, materials to arrive, or equipment to be repaired.

5. *Work-in-Process.* Work-in-process (WIP) includes all stocks, components, and subassemblies in the manufacturing system. The minimum level of WIP is the amount that is being processed (stocks and components that are in a value-added operation or process) at any point in time. Finished goods stored waiting to be invoiced are also counted as WIP.

7. *Defects.* This includes spoiled or rejected parts, subassemblies, finished goods, returns, warranty work, and product recalls.

8. *Scrap.* Material stocks are turned into scrap because of the product design or manufacturing process. Examples include paint overspray, punchings, trimmings, and end-of-reel fragments.

After looking over the types of waste you probably realized that many of the categories don't have any direct environmental impact. However, the indirect impact on the environment can be significant. For example, at a minimum a company that maintains large amounts of WIP needs more floor space, which means a larger building, which means more energy for heat and light. In a lean manufacturing environment a smaller building with lower energy and resource consumption would be just as productive. The excess WIP is a primary source of waste and the associated resource requirements are a secondary waste. In nearly every category of waste it is possible to identify a secondary cost to the environment.

If the waste happens to be in a form that is regulated either directly or indirectly, it will generate additional costs or limitations on a manufacturer. Of course it is essential that companies comply with these government regulations. However, compliance doesn't mean carrying on as before by paying fees and installing abatement equipment. It is essential that manufacturers make choices when they design their product, plan their manufacturing buildings, staff the facility, and select production processes and equipment that will avoid or minimize the cost of the regulated waste.

The Third Source of Waste

The third major source of waste is the materials and activities that are part of the value-added operations and processes used to manufacture the product. To minimize this source of waste requires the designers of the product and the manufacturing engineers and technicians to optimize the materials and processes needed to produce a product. What does optimizing mean? It means selecting materials and processes that minimize waste while providing a product that meets its intended function safely, reliably, and at a value that matches or exceeds its cost to the customer while being profitable for the manufacturer. The phrase "being profitable for the manufacturer" implies being environmentally compliant.

This source of waste is controlled by three groups of people. The first group specifies the product's characteristics and function. The second group uses this specification to design the product and works with the third group to specify how it will be manufactured. These three groups (functions) of people involved in this activity are the following: marketing; product design and engineering; and manufacturing engineering and operations. They are the ones who will select the materials and processes used to manufacture the product.

A basic precept in design states that the selection of a material defines the manufacturing processes that will convert the material stocks into a product. Therefore for a new product the reduction of waste and emissions begins with material selection. Of course someone will be quick to point out that there are several manufacturing processes that can perform a specific operation for a given material. For example, suppose the product design requires that a metal plate must be cut in half. The basic process is called *separating*. There are several ways to carry out this process but we'll consider just three ways; sawing, shearing, and flame cutting. Of these three, shearing would probably result in virtually no material waste, minimal energy use, and no primary emissions.

The methodology for controlling this source of waste uses an approach involving a matrix that considers material classes associated with basic processes. Chapter 2 introduces this approach and provides a tool to evaluate alternatives in materials and processes to identify a design that can be both profitable and environmentally compliant.

A NEW PRODUCT—FIRST PHASE FOR WASTE REDUCTION

Once the costs are obtained the next challenge is to identify alternatives that can reduce these expenses and put the operations into compliance for the least cost. This raises the question, who will do this? We have talked briefly about this in the preceding section on sources of waste. So the answer to this question is; well it depends on whether it is a new product or an existing product—it's a two-stage progression. For a new product the first phase of waste reduction involves marketing, product engineering, and manufacturing engineering to make these decisions. Figure 1.3 shows the groups involved and the inputs guiding their decision making. These inputs, some might call them constraints, form the criteria that shape the design.

How well each group responds to these inputs in defining the product, materials, and processes will determine the potential for waste and its associated costs. Consequently it is critical that they

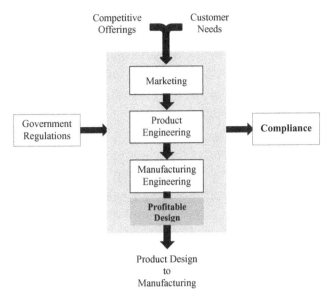

Figure 1.3. The functions and inputs responsible for the design and production of a profitable and compliant product.

recognize the sources of waste and examine alternative designs that meet the customer's requirements. The objective is to create a design that can be produced profitably and be environmentally compliant. Figure 1.3 depicts a linear sequence, which probably never happens. Actually it should never happen. Manufacturing should be consulted and involved in the design of the product just as product design should be involved in marketing—and, marketing should have input in design and manufacturing. This caveat needs to be kept in mind throughout this discussion.

One more comment should be made. A manufacturer that has a product design function (note that many manufacturers do not design the products they make) has to maintain an engineering database that reflects the current state-of-the-art in materials and process technology. The point being that government regulations are continually impacting material stocks and the way they can be processed or used. For example, lead-based paints are no longer included in the engineering database for a furniture manufacturer.

It should be stressed that manufacturing engineering has the responsibility to be aware of the good manufacturing practices that pertain to their industry. Relying only on the processes currently in place will not allow an organization to become or even maintain a competitive and environmentally compliant position in today's marketplace.

One of the best sources for learning about good manufacturing practices is through industry associations and professional organizations. The Society of Manufacturing Engineers (SME) is a professional organization whose goal is to aid manufacturing engineering professionals in staying up-to-date on leading trends and technologies in manufacturing. The American Composites Manufacturers Association (ACMA) is an example of an industry association. This association has over the years carried out practical research on environmentally compliant manufacturing processes and offers education on good manufacturing practices for the composites industry.

Often professional and industry associations provide certification programs for their members, such as the ACMA certification program for composite technicians. The training focuses on the fundamental technologies used in the industry such as open molding (used to make boats and truck bodies), polymer casting, and compression molding. This type of education is one way to develop a foundation for establishing and maintaining profitable and environmentally compliant manufacturing processes.

EXISTING PRODUCTS—SECOND PHASE FOR WASTE REDUCTION

However, the situation changes for plants manufacturing established products using existing processes. The strategy that must be developed in these circumstances will have the same goals—being profitable and in compliance with environmental regulations—but the organizational functions taking the lead in developing this strategy will change from design engineering to production

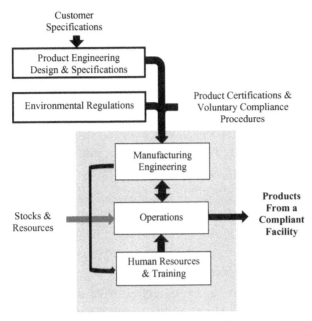

Figure 1.4. The groups involved in reducing waste in an established product design.

and manufacturing engineering and from marketing to human resources and training. Figure 1.4 shows the groups responsible for this stage of manufacturing—producing an established design profitably and in compliance.

These are the people that will take on the responsibility for reducing waste and ensuring that the manufacturing facility is in compliance. If you think about it, this second phase of waste reduction seems to capture more attention because more often product designs tend to be taken as a given. This is true in part because underwriter certifications, life testing, and other forms of product performance qualifications make it more difficult to obtain significant changes in an existing product design. However, there is more latitude in changing processes and operations than there is in changing a stock. But these changes should be done in coordination with the product design group. Nonetheless there are plenty of opportunities for waste reduction at this second stage.

The point is the strategy for environmental compliance should not be limited to abatement, permitting, and lawful waste disposal.

REGENERATION

A factory can be like a basement, garage, or an attic—a place that collects "stuff," stuff that has no immediate use and gets over-looked in day-to-day business. Besides being a symbol for waste, all those bits and pieces get in the way as unneeded fat in a manu-facturing operation. Getting rid of marginal or out-of-date equip-ment and processes gets a manufacturing facility ready to change, adapt, and re-create itself. This also applies to the organization and staffing of the manufacturing facility.

Organizational fat can be detected by observing how an orga-nization handles problems. When a major problem arises does management handle it by creating a new department or group to deal with it? If this is management's approach it makes the problem an ongoing fixed cost causing the organization to swell in size with these specialty groups. Problem solving is a basic function of line management not specialty staffs. Management must clear away its organizational structure as it should clear away out-of-date equip-ment and stocks.

Ideally a plant should be operated in the same way as a conven-tion center or theater for the performing arts. That means it is set up for the event that is currently running but can be quickly changed over to handle a new event that's totally different. Each event will (it must) use profitable and environmentally compliant manufacturing processes. Unfortunately most plants are fixed hard-wired facilities. These facilities have so much inertia that it is nearly impossible to make meaningful improvements quickly or efficiently.

One of the first objections from managers when asked why they are not adopting a more effective method for producing their product is cost. They explain the "price tag" for new equipment and processes makes it prohibitive and they can't afford or don't have the money to invest. This argument misses the point. First of all buying new equipment and trying to squeeze it into some corner of a packed manufacturing floor is not the solution.

The first step is not buying new equipment but getting rid of the waste in the system. Remember the three sources of waste? A comprehensive waste reduction program makes money that later on can be used to update processes or purchase new equipment and locate it where it should be, not where it can be squeezed in. One starting point for regeneration is adopting the concepts and techniques that are known as lean manufacturing. Each of the three sources of waste provides a starting point and the reasons for adopting the values of lean manufacturing.

LIFE CYCLE OF THE MANUFACTURING FACILITY

Life cycle is actually another term for a life span. If you drive around the outskirts of any large city in the Midwest or Northeast one can see the sad sight of huge crumbling factories. But it's not limited to just these areas of the country; during every economic downturn there are factory buildings old and new that go vacant. It is part of the life cycle of manufacturing. Unless an organization is committed to regeneration it will cease to exist. Buildings like people have a finite working life. Companies and organizations are no different unless they become adept at regenerating themselves.

One plant manager explained that a company should never own a building because it is a constraint to its ability to regenerate itself. An apt analogy was the experience of a couple that owned a house that they had lived in for 20 years. An excellent business opportunity became available for them in a town 45 miles away. They took advantage of the opportunity but decided to commute each day instead of moving to the new location. So putting social and family ties aside, the transportation costs and the time spent commuting to avoid relocating became waste caused by the inertia of property ownership.

The unwillingness to be open to regeneration also applies to the people who work in manufacturing plants. Job skills have a finite life too. During most of the twentieth century people with good work habits could find work in factories that paid enough to make unskilled or semiskilled factory work a career. In the twenty-first century that's less likely. A partial explanation is the technological improvements in manufacturing, like the technical revolution that occurred in agriculture a century ago, means there are fewer unskilled and semiskilled factory jobs even though factory output has grown tremendously. Furthermore today's manufacturing jobs require more skills and versatility.

Manufacturers should tailor work so employees have a path to follow that enables them to progress in the organization. This progression is part of the regeneration process. An organization that allows its workforce to "stay put" in one job is committing itself to a finite life cycle. The art and science of current manufacturing has by design created "jobs" that can be mastered in a matter of months if not weeks. After that period of learning the person's job skills too often remain fixed. Persons in these jobs can and do expect increases in pay based on "time on the job." You can probably deduce that in a company like this their wage costs grow not because of innovation or productivity improvements but because of their employee's time in service. Unless the organization can develop a strategy for renewal it will find itself unable to compete against younger companies or companies that are able regenerate themselves.

An example of a company that was faced with this problem of workforce regeneration developed an innovative solution. This company services tooling for the automobile and appliance industry. They repair dies and make replacement punches that fit in the die-sets. These punches create the hole, square, or in some cases the oblong shapes in the metal part that is being stamped out by the die-set. Typically a skilled machinist, a toolmaker, does the work and makes the punches for the die. When the company receives a die-set from a customer they would remove the punches, sharpen the die, make any needed repairs, and replace the punches.

Over the years the company developed an outstanding reputation for their work and meeting tight schedules. However, the company was losing some of its skilled toolmakers through retirement and was even being pressed by competition overseas.

Finding replacements meant trying to hire skilled toolmakers away from other companies. Increasing the pay rate over the standards for the area would only make for more competitive problems. So the company decided on a different course of action. They would make their existing skilled toolmakers mentors and problem solvers. The replacements would be new graduates who had specialized in machinist programs in area community colleges.

One of the many advantages in hiring these graduates was their knowledge of technologies that were not currently being used in the company. These men and women became the initiators that started a regeneration process within the company. The company provided these graduates with good starting pay and a good deal of responsibility for their own projects. With the help and backup from the senior toolmakers these young machinists gained excellent experience in tool and die work. A further benefit, which seemed at first to be a problem, was that these individuals were being lured away by other companies after three or four years on the job. The benefit of course was new graduates came into the company with new enthusiasm and the wage structure remained stable and was not inflated by "length of service" raises. Becoming a "machinist" at this company was a "stepping stone" not a career. If you wanted to remain at this company and have a career you needed to work to become a "senior toolmaker."

CREATING A CLASSIFICATION SYSTEM FOR A COMPLIANT AND PROFITABLE MANUFACTURING SYSTEM

In the next chapter the discussion introduces a systematic approach for evaluating and selecting the materials and processes to satisfy the inputs to the three functions shown in Figure 1.3. The approach will enable engineers, technologists, and managers to systematically analyze the materials and processes used to find opportunities to reduce waste.

To use this approach effectively you'll need to gain an understanding of the scope of the environmental regulations. Therefore in Chapter 3 there is an overview of the federal regulations that govern the creation, handling, and disposal of specific wastes.

In some states and localities there are more stringent regulations and requirements. These state regulations are too region specific and complex to be handled here. Nevertheless the limits and prohibitions that federal regulations establish will serve to expand the definition of waste. Also, the federal regulations do provide a framework that can help you understand the approach used by individual states and localities. It is particularly important to factor in the impact that these regulations will have at the design stage when material stocks and processes are being evaluated and compared.

In either event, designing from new or manufacturing an existing product, the material and process classification system is a basis for creating a strategy for being a profitable and compliant manufacturer. Once the materials and processes are established there are a multitude of tools that will be needed to implement a strategy to become a cost-effective "green" manufacturing facility. Many of these tools are introduced in the case studies in Chapter 4. The introduction to these tools or methods and techniques then continues in Chapter 5.

Overall the sequence of chapters as presented follows a path that can be used to establish and implement an effective strategy. However, as with most endeavors in manufacturing this will be an ongoing activity—once again think about Ford's fourth principle. To be successful the strategy developed should be a means to create a way of thinking, behaving, and evaluating. The tools and methods are activities. The outcomes for the strategy being followed should be profitability, compliance, and the ability to regenerate the manufacturing facility.

SUMMARY

Nearly a hundred years ago Henry Ford and key members of his company started work on a new design. It was not a car but a manufacturing system. The impact it had on the country and the company was revolutionary. It made his company profitable and his company's product affordable for nearly every working family in the country. Ford's manufacturing system became a pathfinder for industry in the United States. The vertically integrated manufacturing system he created enabled his company to capture nearly all the valued-added opportunities in the sequence of manufacturing. This strategy worked well as technology evolved during the first half of the twentieth century. Today product technology is changing much more rapidly. The vertically integrated manufacturing operation has a difficult time in regenerating itself so that it can adopt new know-how quickly and profitably.

Today manufacturing is facing the same challenges that Ford faced—making products that are affordable and profitable. However, part of this affordability involves reducing the environmental costs of production. Companies today can gauge these costs by tracing their effectiveness in transforming materials and resources into finished goods.

As the twentieth century manufacturers realized, each step in the manufacturing sequence provides an opportunity to profit in proportion to the amount of value added at each step. However, each step carries costs that include the environmental impact of processes and operations. The strategies that can be employed to control these costs may be as simple as cost avoidance. Companies can choose to limit their involvement to just one or two steps of the manufacturing sequence. The most extreme and simplistic strategy would be for a "manufacturer" to design a product and have others produce it. The manufacturer, in name only, would be involved in just one step of the manufacturing sequence, the product's distribution. The most complex strategy would be Henry Ford's approach in creating a manufacturing operation that spans nearly the entire manufacturing sequence.

The life cycle analysis (LCA) that came into use nearly 40 years ago is an ideal technique for cataloging and assessing the transformation of resources into products and the environmental impact. This analysis involves all three stages of the manufacturing sequence.

- Stage 1, creating the manufacturing stock
- Stage 2, product manufacture
- Stage 3, distribution sales & service and disposal.

However, any one company that would attempt to span the entire manufacturing sequence would require a great deal of technical and management skill to effectively control all the costs, which includes environmental compliance. This can be done. In fact being "Green" can be the basis of a strategy for improving a company's profitability. Both green manufacturing and lean manufacturing share a common goal, to get rid of waste. Waste includes everything that does not add value to the product in the hands of the buyer. Eliminating waste is the foundation for creating a profitable manufacturing operation.

There are two groups of people directly responsible for eliminating or minimizing waste. The first group is involved when the product is being designed and all options are available to create a "green and lean" product design. The second group takes over this responsibility when the facility is producing an established design. There the options to change material stocks and processes are limited. Nonetheless there are still opportunities to reduce waste.

Many of these opportunities arise if the company is able to regenerate, adopt change, and renew itself. Companies that are rigidly structured both organizationally and physically limit or impede their ability to sustain themselves. This resistance or inability to regenerate creates waste, which means it can be measured as a cost. Consequently an organization needs to adopt a systematic approach for identifying and quantifying the cost of waste as a basis for designing a production system that is compliant and profitable.

SELECTED BIBLIOGRAPHY

American Composites Manufacturers Association. Available at http://www.acmanet.org/ (accessed January 26, 2012).

Ford, H. (1922). *My Life and Work.* Kindle Edition, 2005. (Originally published in 1922 by Garden City Publishing Co., Garden City, NY.)

Joint Research Centre. Life cycle thinking and assessment. Available at http://lct.jrc.ec.europa.eu/ (accessed January 8, 2012).

Michalski, W., and King, D. (2003). *Six Sigma Navigator.* New York: Productivity Press.

New York Times. (2009). Value of goods. Available at http://www.nytimes.com/2009/02/20/business/worldbusiness/20iht-wbmake.1.20332814.html (accessed January 8, 2012).

Society of Manufacturing Engineers. Available at http://www.sme.org (accessed January 26, 2012).

Standard, C., and Davis, D. (1999). *Running Today's Factory.* Dearborn, MI: Society of Manufacturing Engineers.

United Nations Statistics Division. Manufacturing value. Available at http://unstats.un.org/unsd/snaama/cList.asp (accessed (January 10, 2012).

U.S. DoE. Federal Energy Management Program, Building Life-Cycle Cost (BLCC) Programs. Washington, D.C.: U.S. Department of Energy, Office of Energy Efficiency and Renewable Energy. Available at http://www1.eere.energy.gov/femp/information/download_blcc.html#blcc (accessed January 4, 2012).

U.S. Environmental Protection Agency. Life cycle analysis. Available at http://www.epa.gov/nrmrl/lcaccess/ (accessed January 8, 2012).

CHAPTER 2

BUILDING A DECISION-MAKING MODEL

INTRODUCTION

In the first chapter we described manufacturing in general and the ways companies organize their manufacturing operations to create high-quality products. As you can well imagine, manufacturing includes virtually every endeavor where tools are used to transform materials into hard-good consumer products. Here in Chapter 2 we present a systems model of manufacturing. We use an approach that will enable us to discuss the different parts of the system independently before putting them all together in a results-oriented tool we call the *profitable and compliant process chart* (PCPC). You will soon understand how the PCPC will help you make appropriate choices about the selection and use of materials and processes to create products, components, or subassemblies that are both profitable and environmentally compliant.

Improving Profitability Through Green Manufacturing: Creating a Profitable and Environmentally Compliant Manufacturing Facility, First Edition.
David R. Hillis and J. Barry DuVall.
© 2012 John Wiley & Sons, Inc. Published 2012 by John Wiley & Sons, Inc.

At first glance trying to model manufacturing may appear to be quite complex. That is because of our natural tendency to look for a quick and easy fix for accomplishing process improvement goals. Because of our impatience we often rush to make decisions about all of the parts of a complex system at the same time. This may result in poor decision making, confusion, and inefficiency.

The approach we use here will work equally well with manufacturing industries that are very different from each other in terms of the products they produce and the materials and processes used to manufacture these products.

Now it is time to get started. The next few sections describe some of the methods used to identify and classify industries. This will help in the development of a catalog of materials and processes that will be needed later in this chapter. Once you have gone through this step-by-step process you will be able to create your own catalog or taxonomy for any manufacturing industry.

INDUSTRIAL PRODUCTION AND MANUFACTURING

There are several different systems used to classify manufacturing firms for analysis. The *Industry Classification Benchmark* is a system that is popular in the fields of finance and market research. It was developed by Dow Jones and the FTSE Group (often known by its nickname "the Footsie"). Another is the *Global Industry Classification Standard*, a major system for equities developed jointly by Morgan Stanley Capital International (MSCI) and Standard & Poor's. However, the *North American Industry Classification System* is the most significant system used in North America to classify industry. Each of these classification systems groups industries according to similar functions and markets and identifies the types of companies producing related products. There are other systems that are less formal but are discipline specific. That is, they are used by specific types of industries and are organized according to products: for example, manufacturing industry, chemical industry, electronics industry, pharmaceutical industry, automotive industry.

The North American Industry Classification System (NAICS), pronounced *Nakes*, was adopted on April 9, 1997, and was developed by the Office of Management and Budget (OMB) in collaboration with Canada's U.S. Economic Classification Policy Committee (ECPC), Statistics Canada, and Mexico's Instituto Nacional de Estadística y Geografía. NAICS provides a solid foundation for creating a model or taxonomy of manufacturing. The NAICS system is used by statistical agencies to codify business and industrial establishments when collecting economic data. It is also used for other, nonstatistical purposes: administrative, regulatory, contracting, and taxation. Often contracting agencies require firms to register their NAICS codes to determine their eligibility to bid on contracts. NAICS is the guidebook used by business, industry, and governmental agencies to classify industry in the United States, Canada, and Mexico. It should be kept in mind that NAICS is an *industry* classification system, not a product classification system. Another system, called the *North American Product Classification System* (NAPCS), is currently being developed to integrate all of the industries defined in NAICS in terms of their products. The emphasis will be on goods produced and services provided.

NAICS was built on a solid foundation established by its predecessor, the *Standard Industrial Classification Index* (SIC), which in 1939 became the first standard industrial classification for the United States. The first List of Industries for manufacturing in the United States was published in 1938. This list was consolidated with the 1939 List of Industries for nonmanufacturing industries to become the SIC.

For hundreds of years manufacturing establishments have developed and refined best practices for using tools, manufacturing processes, production techniques, machines, and manufacturing systems for the production of products. *Industrial production* is a term used by the Federal Reserve Board (FRB) to refer to the total output of U.S. factories and mines and is a key economic indicator for the U.S. economy. The National Association of Manufacturers (NAM) states that "the United States is the world's largest manufacturing economy, producing 21 percent of global manufactured products" (NAM, 2011). Manufacturing is responsible for two-thirds of all of the research and development in this nation.

Industrial production is accomplished by transforming raw materials into industrial *stocks* such as sheets, bars, wire, rods, pellets, powders, and liquids that are used to make new products or remanufacture old products. The process of manufacturing begins when a *primary manufacturing industry* produces the first form of stock. Let's use the china clay industry as an example to understand the complexity of stock production. Manufacturing begins when the primary manufacturing industry, a china clay (kaolin) mine, extracts the clay from the ground. The run-of-mines (ROM) clay is then beneficiated to remove mineral impurities, get the desired particle size, and obtain the needed physical and optical properties. Here, beneficiating is the process of removing minerals from the extracted ore, normally using processes such as crushing, grinding, and froth flotation. All of these processes are part of what we refer to as *Stage 1* manufacturing. A variety of manufacturing processes are normally involved: size separation using levigation (washing), wet hydrocycloning, magnetic separation, froth flotation, acid leaching or bleaching, drying, micronizing, and air-classification surface treatment. The resulting clay stock that is created is a dry powder that is bagged and sold as a stock to *Stage 2* manufacturing industries. In most cases these would be china potteries that convert the stock into products such as china plates, cups, tiles, figurines, paints, and industrial coatings. *Stage 3* industries then concentrate on product distribution, sales, and service and on disposal of the worn-out products.

The concept of stages of industrial production is often referred to as the *three-sector hypothesis* (Fourastié, 1949). Fourastié separated industrial activity into three sectors, or stages as they are called here. The first sector, or Stage 1, includes primary manufacturing firms such as agriculture, agribusiness, fishing, aquaculture, forestry, and all mining and quarrying. In many cases other types of manufacturing companies located in close proximity to the primary companies will collect, pack, package, refine, or process the raw materials that are processed by this sector. Next Fourastié referred to a secondary sector, or Stage 2 activity, that produces products. The tertiary or third sector, Stage 3, includes the distribution and service of the manufactured products,

When most people think about manufacturing they think about clusters of medium- or small-sized factories in an industrial park. There are other types of establishments as well: steel mills, lumber mills, paper mills, or cotton mills that are often located close to the source of their raw materials. But manufacturing doesn't have to be done in a facility such as a factory or mill. Production strategies such as agile manufacturing, mass customization, lean manufacturing, and reconfigurable manufacturing are practiced today, creating customized products in varied locations. Approaches such as these require increased flexibility, reduced setup times, and short delivery times. These requirements provide opportunities for entrepreneurs who are more mobile. Some of these mobile manufacturing work cells are in trailers or vans that are pulled or driven to a job site. Mobile manufacturers often provide specialized manufacturing on demand to work areas within a plant or to subcontractors in temporary locations. This is manufacturing on demand, serving new markets with local production.

It also enables sharing of specialized equipment that may be underutilized by one company and needed by another. Mobile factories may also provide a backup and measure of security when a quick response to peaks in production is needed. You may have come across mobile work cells operating in your area: mobile welders, portable sawmills, seamless-gutter fabricators, mobile locksmiths, mobile tube benders, or mobile machining centers.

No matter where manufacturing occurs, in a plant, in a garage, or at a remote job site, the basic steps are always similar. The input is an industrial stock and the output is a manufactured part or product.

An important part of today's world of manufacturing involves the assembly of processed materials or stock produced by other companies. In many cases stock is produced in industries outside of what we are classifying as manufacturing. Some examples are agriculture, forestry, fishing, mining, and quarrying.

CLASSIFYING MANUFACTURING INDUSTRIES

Let's take a closer look at how NAICS defines manufacturing and classifies manufacturing industries. It is important to understand that the boundaries between different industries and subindustries are not precise and can be somewhat blurry. Nevertheless, this system will help us understand the magnitude and complexity of manufacturing and its importance to global economic development and conservation of natural resources. It is important to take time to understand what manufacturing is all about before making choices about materials and processes. NAICS is an essential tool used by industry, business, and governmental agencies.

According to NAICS, manufacturing companies are those firms that are involved in the "mechanical, physical, or chemical transformation of materials, substances, or components into new products" (OMB, 2010). The assembly of component parts is also classified as manufacturing. This does not include activities that would be classified in Sector 23, Construction. The following are examples of activities considered to be manufacturing: fabricating metals; wood preserving; heat treating and polishing for the trade; apparel jobbing; printing and related activities; water bottling and processing; milk bottling and pasteurizing; fresh fish packaging (e.g., oyster shucking, fish filleting); producing ready-mixed concrete; leather converting; grinding of lenses to prescription; electroplating, plating, and lapidary work for the trade; fabricating signs and advertising displays; rebuilding or remanufacturing machinery (i.e., automotive parts); ship repair and renovation; machine shops; and tire retreading.

NAICS uses codes 31–33 to describe the primary manufacturing sectors. Each of these sectors is broken down further according to the type of establishment. For example, code 313 is assigned to textile mills. This is broken down further: fiber, yarn, and thread mills (code 3131); fabric mills (code 3132); and textile and fabric finishing, and fabric coating mills (code 3133). The four-digit numbers refer to specific types of establishments.

There are 21 *major product groups* (MPGs) for manufacturing listed under codes 31–33. This is where we will begin our journey. The 21 MPGs from NAICS cover 478 different six-digit subclassifications identifying types of manufacturing establishments. The MPGs for manufacturing are shown in Table 2.1.

Let's take a moment to think about these classifications. From the perspective of an economist we might describe the output of these industrial classifications in terms of products that are "durable goods" or "nondurable goods." Durable goods are also called *hard goods*. Nondurable goods are referred to as *soft goods*.

TABLE 2.1. Manufacturing Product Groups in NAICS 2011

Code	Type of Manufacturing	Code	Type of Manufacturing
311	Food manufacturing	325	Chemical manufacturing
312	Beverage and tobacco product manufacturing	326	Plastics and rubber products
		327	Nonmetallic mineral products
		331	Primary metal manufacturing
313	Textile mills	332	Fabricated metal products
314	Textile product mills	333	Machinery manufacturing
315	Apparel manufacturing	334	Computer and electronic product manufacturing
316	Leather and allied product manufacturing	335	Electrical equipment, appliance, and component manufacturing
321	Wood product manufacturing	336	Transportation equipment manufacturing
322	Paper manufacturing		
323	Printing and related support activities	337	Furniture and related product manufacturing
324	Petroleum and coal products manufacturing	339	Miscellaneous manufacturing

A durable good is a hard good that lasts for three or more years and continues to provide value or be useful throughout its life cycle. It is a product that is not completely consumed in use. Examples of durable goods are products such as jewelry, cast iron frying pans, and bricks because in most cases they never wear out. Goods such as cars, refrigerators, furniture, and air conditioning units are also durable hard goods lasting longer than three years, but they do eventually wear out. Other examples of durable hard goods are toys, lawn mowers, consumer electronics, and a long list of goods that we use at home and at work.

Nondurable goods are products that are immediately consumed, wear out, or have a life cycle of less than three years. Examples of nondurable consumer goods include food, household cleaning products, cosmetics, packaging, textiles, paper products, office supplies, pharmaceutical products, tooth brushes, and fuel.

MAJOR PRODUCT GROUPS FROM NAICS

You will discover many similarities and some differences within the MPGs in terms of materials, processes, skills, and techniques that manufacturers use to create their products. It would be difficult to analyze all of these MPGs and develop a single catalog or taxonomy for conducting a "profitable and compliant manufacturing analysis."

Our approach concentrates on durable or hard good consumer products when creating our foundation for the *profitable and compliant process chart*. Therefore we need to identify those manufacturing sectors listed in NAICS that do not generate hard good products with a life cycle of more than three years. These sectors are shown in Table 2.2 (manufacturing sectors disqualified from classification) and are not included in our developmental process.

After removing the 10 groups shown in Table 2.2 we are left with 11 MPGs from the NAICS classification. However, a careful look at these reveals commonalities in terms of materials and

TABLE 2.2. Manufacturing Sectors Disqualified from Classification[a]

Code	Type of Manufacturing	Code	Type of Manufacturing
312	Beverage and tobacco	322	Paper
313	Textile mills	323	Printing and related support
314	Textile product mills	324	Petroleum and related
315	Apparel		products
316	Leather and allied products	325	Chemical

[a]That is, sectors that do not meet criteria for inclusion in our classification.

processes. Sector 331, Primary Metal Manufacturing, Sector 332, Fabricated Metal Product Manufacturing, and Sector 333, Machinery Manufacturing, are each concerned with products made primarily of metal and use similar manufacturing processes. Also, a closer look at Sector 339, Miscellaneous Manufacturing, reveals that this area is responsible for the manufacture of medical equipment and supplies. This sector also relies heavily on metal materials. Sector 336, Transportation Equipment, also has a high concentration of metal stocks in its products. For our analysis we group all of these sectors together and refer to them as *metal product manufacturing*.

A review of Sector 327, Nonmetallic Mineral Product Manufacturing, shows that the major material involved in this sector is clay products used in the ceramics industry. We label this as *ceramic product manufacturing*.

Sector 321, Wood Product Manufacturing, and Sector 337, Furniture Product Manufacturing, are each concerned with products made primarily of wood and share common manufacturing processes. We combine these sectors and group them under *wood product manufacturing*.

Two other sectors also need to be analyzed more carefully—Sector 334, Computer and Electronic Product Manufacturing, and Sector 335, Electrical Equipment, Appliance, and Component Manufacturing, which emphasize many different types of materials in their products. It is not practical to classify them in terms of their basic material because they do not have one. Products in

TABLE 2.3. Major Product Groups (MPGs)

Industry Sector	Code	Type of Products
Wood product manufacturing	321	Wood products
	337	Furniture products
Ceramic product manufacturing	327	Nonmetallic mineral products
	334	Computer and electronic products
	335	Electrical equipment, appliance, and component products
Plastic and rubber product manufacturing	326	Plastic and rubber products
	334	Computer and electronic products
	335	Electrical equipment, appliance, and component products
Metal product manufacturing	331	Primary metal products
	332	Fabricated metal products
	333	Machinery manufacturing
	334	Computer and electronic products
	335	Electrical equipment, appliance, and component products
	336	Transportation equipment
	339	Miscellaneous manufacturing

these sectors have many different types of materials, components, and subassemblies and rely heavily on products from the other sectors. We address content from these industries under our other classification rubrics.

At this point four MPGs remain and are part of our decision-making model (refer to Table 2.3). These MPGs were selected because the companies in these groupings use major materials and have common manufacturing processes to make their products. They provide the foundation we use to build a PCPC for environmentally conscious green manufacturing.

The next step in the classification process is to look closer at these four MPGs in terms of their enterprises or *companies* and products. We refer to the manufacturing companies as *major types of establishments* (MTEs). Stop for a moment and think about the MTEs and how stock and products are generated from wood materials. Refer to Table 2.4.

First, loggers (Stage 1 manufacturers) cut the trees and provide these logs to sawmills, where boards, dimension lumber, and many

TABLE 2.4. Major Types of Establishments and Their Products—Woods

Stage 1			Stage 2	
MTE	Stock		MTE	Products
Loggers	Debarked and trimmed logs (stock for sawmill)			
Sawmills	Boards, dimension lumber, beams, timbers, poles, shingles, siding		Furniture manufacturing	Household and institutional furniture, kitchen cabinets and countertops, upholstered household furniture, institutional furniture, cabinets and office furniture
			Woodwork and millwork manufacturing	Custom architectural woodwork and millwork
			Showcase, partition, and shelving manufacturing	Showcase, partitions, shelving, mattresses
			Blinds, shades, and fixtures manufacturing	Blinds, shades, fixtures
			Furniture parts and frames manufacturing	Furniture parts and frames
			Multidimension fiberboard manufacturing	Fiberboard, insulation board, block board
			Engineered wood manufacturing	Veneer, sheeting, compreg, impreg
			Pulpwood manufacturing	Wood chips, shavings, fiberboard, bark
			Wood treating and preserving manufacturing	Posts, pilings, flooring, millwork, and structural lumber treated to prevent rotting
			Construction: manufacturing of products and components	Residential and commercial structures, trusses, building materials, prefabricated manufactured construction components

MTE, major type of establishment.

43

other types of products are created. These products are used as stocks by Stage 2 manufacturers to make their own products—things such as kitchen cabinets, building materials, boats, furniture, toys, crates, pallets, barrels, and engineered wood materials. Many wood stocks are used by industries outside of manufacturing to make their own products. Construction is an example.

Most of the MTEs that use wood stocks are engaged in one or more of the following types of activities: (1) sawing or shaping dimension lumber, boards, beams, timber, poles, ties, shingles, shakes, siding, and wood chips created from logs; (2) treating sawed, planed, or shaped wood with preservatives to prevent decay and to protect against fire and insects; (3) creating engineered wood products using wood and other materials; (4) using stock created by others to make products.

The final product produced by these first-level enterprises is sometimes considered a product by consumers: for example, wood shavings used for dog bedding or wood chips used to cover flower beds. Often the production of a stock needs only a minor amount of additional processing to create a product: for example, fence posts. The only additional processing required is impregnating the post with a preservative to protect it from insects or rotting. This can be done by the Stage 1 facility and sold by them as a product to a consumer.

There are other circumstances that can blur the distinction between Stage 1 (primary) and Stage 2 (secondary) manufacturing. Sometimes more processing than a preservation treatment is needed before the stock is useful in the final product. would be Plywood, the stock used to make wooden cabinets, is an example. In this situation thin sheets of veneer are laminated to create sheets of plywood for the cabinet. The plywood "sandwich" is created using the veneer sheets, adhesives, pressure, and heat. This processing can be done by a secondary manufacturer that will sell "cabinet grade" plywood panels as stocks to other manufacturers.

There are often many successive transformations of stock into different forms of stocks before the final product is manufactured. Sometimes one company's final product is a stock for another

TABLE 2.5. Major Types of Establishments and Their Products—Ceramics

Stage 1		Stage 2	
MTE	Stock	MTE	Product
Sand quarry	Silica sand	Brick manufacturing	Bricks
		Glass manufacturing	Glass and glass products
		Semiconductor and component manufacturing	Semiconductor and electronic components
Gravel pit	Gravel, crushed rock, and stone	Construction and roadway manufacturing	Concrete for construction, for mixing with asphalt, as construction aggregates
		Industrial material manufacturing	Concrete blocks, bricks, pipes, roofing shingles, abrasives
Stone quarry	Stone	Stone veneer and tile manufacturing	Stone products, burial monuments and statues, tile, and veneer
Clay quarry	Clay	Dinnerware and pottery manufacturing	Dinnerware, pottery, ceramic products
Refractory quarry	Quartz, bauxite, quarry stone	Refractory product manufacturing	Containers, concrete products, lime, gypsum, and abrasives

MTE, major type of establishment.

manufacturer. Nevertheless it is still necessary to recognize the distinction between primary Stage 1 and secondary Stage 2 manufacturing.

Let's review the MTEs and MPGs making *ceramic* products (see Table 2.5).

In Table 2.5 you reviewed the MPGs and MTEs that utilize ceramic stocks to make their products. Stage 1 companies were classified by the type of material involved: sand, gravel, stone, clay, or refractories. The refractory industry is concerned with high-temperature applications of sand, gravel, stone, and clay. Stage 2 MTEs are classified by product type.

Let's take a moment now to look at another MPG, plastics and elastomerics (rubber). In Table 2.6 you see that Stage 1 manufacturers are classified by type of plastic material: thermoplastics;

TABLE 2.6. Major Types of Establishments and Their Products—Plastic and Rubber

Stage 1		Stage 2
MTE	Stock	MTE	Product

Stage 1 MTE	Stock	MTE	Stage 2 Product
Chemical plant: thermoplastics	Polyethylene	Plastic bag manufacturing	Shopping bags
	High-density polyethylene	Plastic package manufacturing	Milk jugs, detergent bottles
	Low-density polyethylene	Tile manufacturing	Floor tiles, siding
	Polyethylene terephthalate	Packaging material manufacturing	Drink bottles, jars, plastic film, microwavable packaging
	Polypropylene	Plastic product manufacturing	Bottle caps, drinking straws, appliances, car bumpers, plastic pressure pipes
	Polystyrene	Foam manufacturing	Packaging "peanuts," tableware, CD boxes
	Polyvinyl chloride	Laminated plastic plate, sheet, and shape manufacturing	Plumbing pipe, window frames, shower curtains, flooring
	Polytetrafluoroethylene	Plastic product manufacturing	Plumbing thread seal tape, bearings
	Polyester fiberglass	Sheet and bulk molding compound manufacturing	Automotive exterior panels; radio and appliance knobs
	Urea-formaldehyde	Urethane and other foam product manufacturing	Particleboard, medium-density fiberboard, foamboard
	Epoxy resin	Composite boat manufacturing	Fiber-reinforced composites, boats
	Melamine resin	Plastic product manufacturing	Countertops, dry erase boards, glues

	Material	Manufacturing	Products
Chemical plant: thermosets	Polyester	Automotive fabric manufacturing	Textiles and fibers
	Polyester fiberglass, sheet and bulk molding	Boat manufacturing; automotive panel manufacturing; knob and billet manufacturing	Boats, marine canopies, automotive panels
	Catalysts and additives	Catalyst manufacturing	
	Urea-formaldehyde foam	Urethane foam product manufacturing	Thermosetting parts Medium-density fiberboard, and foam board
	Epoxy resin	Composite boat manufacturing	Fiber- and glass-reinforced plastics, boats
	Melamine resin	Laminated plastic plate, sheet, and shape manufacturing	Countertops, dry erase boards, glues, flame retardants
	Polyimide	Semiconductor and other electronic component manufacturing	Printed circuit boards, high performance industrial tape
	Nylon	Plastic product manufacturing	Fiber, fishing line, brushes for toothbrushes
	Acrylonitrile butadiene styrene	Plastic pipe and pipe fitting manufacturing	Drainage pipe, computer equipment, electronic accessory cases
	Polycarbonate	Polycarbonate resin manufacturing	Traffic lights, eye glasses, CDs, lenses
	Polyurethane	Urethane product manufacturing	Car parts, seat cushions, coatings, rollers
	Phenol-formaldehyde resin (Bakelite)	Resin compounding manufacturing	Electrical insulators and plastic wear handles
Chemical plant: elastomers	Synthetic rubber: latex	Synthetic rubber manufacturing	Tires, ties, hoses, and belting

thermoset plastics; and elastomerics. Stage 2 manufacturers are classified by their products.

Three Stage 1 chemical plants are shown in Table 2.6. Scientists, engineers, and technicians in these plants prepare different types of plastic resins and additives. Plastic materials, which include elastomerics (synthetic and organic rubber), are all made from hydrocarbon polymers. There are many different types of thermoplastics and thermoset plastics, and almost 20 different chemical types and grades of elastomers (synthetic rubber).

The last MPG to be reviewed is "Metals" (refer to Table 2.7). This is an important MPG that is a major indicator of economic progress and prosperity. According to the U.S. Geological Survey and Conference Board Report for May and June 2011, the "primary metals" sector was concerned with almost 25 percent of the sales of domestic products in the United States (USGS, 2011a). Recycled metal materials are used for a diverse array of products, ranging from cans, windows, doors, and file cabinets to bridges, fire hydrants, and utility poles. In 2010, 41 percent of the aluminum scrap that was purchased came from recycled aluminum products (USGS, 2011b). The price of metal is viewed by many global investors, traders, and forecasters as an indicator of economic progress. Metal prices are influenced by demand for the products from which they are made. World-renowned trader Dennis Gartman watches the prices of copper, steel, and aluminum indexes to judge consumer interest in making purchases throughout the world. According to Gartman the price of many base metals will drop long before economic data indicate a weakness in the economy. He used these metal indexes to forecast the 2008 slump (Gartman, 2009).

Two types of Stage 1 industries are shown for mills and primary metal manufacturing in Table 2.7. The MTEs shown for primary metals are classified by their products. Five different manufacturing sectors from the NAICS classification system are included in metals: primary metals; fabricated metals; machinery manufacturing; transportation equipment; and medical equipment. The major material used in each of these sectors is metal.

TABLE 2.7. Major Types of Establishments and Their Products—Metals

Stage 1		Stage 2	
MTE	Stock	MTE	Product
Iron and steel mills	Iron ore	Ferroalloy manufacturing Steel product manufacturing	Steel products, iron products
		Iron foundries	Powdered metal products
		Cutlery and hand tool manufacturing	Cutlery and hand tools
		Saw blades and handsaw manufacturing	Saw blades and handsaws
		Metal can, box, and container manufacturing	Boxes and containers
Copper mill	Copper ore	Copper and tin smiths Copper wire drawing Copper rolling, extruding, drawing	Ornamental copper products, roofing
Alumina refining mill	Aluminum and other nonferrous metals	Aluminum, nonferrous, and primary aluminum manufacturing	Roofing, siding, structural support, lightweight beams and trusses
Smelters, refineries, rolling, drawing mills	Beneficiated ore	Metal stamping manufacturing	Fence, panels, siding, wire coil, stampings, wire forms, retaining rings
	Iron	Ferrous metal foundries	Metal castings
	Hot/cold rolled steel	Roll forming manufacturing	Rings, struts, angles, and stampings
	Hardened steel	Spring manufacturing	Springs
	Hot/cold rolled steel	Metal heat treating	Screwdriver tips, chisels, and other hardened products
Primary metal manufacturing: electrical	Copper, aluminum, tungsten wire	Electrical lighting equipment manufacturers	Other electrical lighting equipment and components
	Sheet, cast, and machined metals	Household appliance manufacturing	Refrigerators, microwave ovens, and waste compactors
	Sheet, cast, machined metals, wire, and components	Electrical equipment manufacturing	Electrical equipment
	Sheet, cast, machined metals, wire, and components	Computer and peripheral equipment manufacturing	Computer and peripheral equipment

(*Continued*)

49

TABLE 2.7. (*Continued*)

Stage 1		Stage 2	
MTE	Stock	MTE	Product
Primary metal manufacturing; computer and electronic	Sheet, cast, machined metals, wire, and components	Communications equipment manufacturing	Communications equipment
	Sheet, cast, machined metals, wire, and components	Electronic computer manufacturing	Computers
	Sheet, cast, machined metals, wire, and components	Audio and video equipment manufacturing	Audio and video equipment
	Sheet, cast, machined metals, wire, and components	Radio and TV broadcasting equipment	Radio and TV broadcasting equipment
	Silicon, germanium, aluminum nitride, gallium antimonide, cadmium sulfide, zinc oxide	Semiconductor materials manufacturing, other electronic components and products, superconductor manufacturing	Semiconductors, piezoelectric components, radar detection diodes, infrared detectors, photoresistors, solar cells, window coatings, superconductors
	Copper	Printed circuit board manufacturing	Printed circuit boards
	Copper and aluminum wire	Electronic component manufacturing	Electronic components
	Metal oxide Nichrome wire	Electronic resistor manufacturing	Resistors, potentiometer, cermets
	Oxide formed on an aluminum or tantalum foil	Electronic capacitor manufacturing	Capacitors
	Sheet, cast, machined metals and components	Navigational, measuring, electromedical, and control instruments manufacturing	Navigational, measuring, electromedical, and control instruments
	Sheet, cast, machined metals and components	Manufacturing and reproducing magnetic and optical media manufacturing	Magnetic and optical media

MATERIAL STOCKS

Now you should have a pretty good understanding of manu-facturing, its MPGs and MTEs. We are ready to shift our focus to manufacturing materials and processes. In Chapter 1 the term "manufacturing sequence" was introduced to describe the sequence or steps in the creation of a product. We know that for all MPGs the manufacture of a product starts when raw materials are extracted and converted into industrial stocks by primary manufacturing industries. Next MTEs purchase these stocks, apply manufacturing processes, and use assembly operations, tools, and techniques to transform these materials into new products. These secondary manufacturing industries are Stage 2 manufacturers. Once a product is completed it enters the tertiary sector, or Stage 3 manufacturers, to be distributed, sold, serviced, and then dis-posed of.

You may wonder just how it can be possible to make sense of all of the MPGs, MTEs, and thousands of materials and combina-tions of materials and stocks that are used to make hard good consumer products. The answer is simple. We use the concept of systems analysis to organize our thinking and reduce the scope of coverage.

There are many options available when it comes to selecting what is believed to be the best material for making a specific product. Often the material of choice is based on tradition, the fact that it has been used for a long time, not necessarily because it is the best choice. Some materials are less expensive than others, and they may or may not be easier to process. Also all materials have an impact on the environment when they are used in manu-facturing. These are the factors that influence cost, which is another way of saying waste.

Remember that the purpose of this book is to provide a method that can be used to analyze the impact of materials and processes used in making products, and to advance strategies for reducing waste in all stages of production. This will be important not only if you are designing new products but also if you are improving or reworking existing products.

MAJOR MATERIAL FAMILIES

We classify materials using the concept of *major material families* (MMFs). In the preceding tables you were introduced to many of the materials that will be included in the MMFs. You also read about *major manufacturing enterprises* and their products made from wood, metal, ceramic, and plastic materials. We view these families of material as foundational. However, there is one more material family that we haven't talked about yet—composites. Composites are a stock that is created when two or more different but complementary substances are combined to create the final material. Composites contain a filler, normally fibers, particulates, or fabric, suspended in a matrix binder. Examples include polymeric (plastic) matrix composites, metal matrix composites, and ceramic matrix composites.

Most of the industrial stock that is used by manufacturing industries and the finished products that are created are made from these five classes of materials. There are exceptions to the rule, but we are not focusing on exceptions. To implement a PCPC that generates positive outcomes we must concentrate on what is most important.

When a material is selected the first consideration is how to transform the material stock into the designed part or component using one or more manufacturing processes. Selecting the most appropriate material for a manufactured product is sometimes a challenging task. Often the choice of the material is limited by the processes that are available in the manufacturing facility. Let's look closer at plastics and metals to understand just how complicated this can be.

For 35 years plastic has been the most widely used material in the United States. Few products can be found today that do not include some plastic parts or components. Today there are hundreds of varieties of plastics being produced but there are only three classes of plastics: thermoplastics, thermosets, and elastomerics. The behavior of these classes of plastics is very different from each other.

Shift your attention to metals. There are more than 200 different copper alloys, each with its own classification number. There are thousands of different types of steel. Steel is an alloy of iron and carbon, but other materials may be included in the alloy, such as such as manganese, chromium, nickel, molybdenum, copper, tungsten, cobalt, vanadium, or silicon. The percentages of these materials in the mixture are what influence the hardness, ductility, corrosion resistance, and other unique characteristics of steel.

Add to this the many different types of materials that might be available in the other MMFs: wood, ceramics, and composites. There are many materials at our disposal and each of these provides challenges in terms of what happens to the environment when used in production. Some materials are more opportunistic than others.

Refer to Table 2.8 and take a moment to study it carefully, thinking about the products made from these materials in your own interest area.

BASIC PROCESS CLASSIFICATIONS

Secondary manufacturing firms have common characteristics related to their materials and processes, tools, techniques, and systems. Some of these characteristics were covered in Chapter 1 and others here in this chapter when we discussed major manufacturing enterprises. But there is more to this story. You know that secondary manufacturing firms are those that transform industrial stock, using manufacturing processes, into hard and soft good products, components, and subassemblies.

The missing link is manufacturing processes. Just as we did before when organizing materials for analysis, we need to simplify our approach for identifying and selecting the most appropriate

TABLE 2.8. Major Material Families and Their Material Stocks

MMF	Types	Material Stock
Wood	Hardwoods	Oak, walnut, maple, birch, ash, cherry, mahogany, poplar, teak, elm
	Softwoods	Cyprus , pine, cedar, fir, hemlock, redwood, spruce
Ceramic	Traditional	Clay: refractories, cement, tile, block, brick, concrete, plaster, abrasives
		Glass: float glass, E-glass, cast glass, flat glass
	Industrial, engineered, or advanced ceramics	High-performance oxide, aluminum oxide, aluminum titanate, piezoceramics, silicate ceramics, zirconium oxide, silicon carbide, silicon nitride, aluminum nitride, metal matrix composites, ceramic matrix composites, electro ceramics, braze alloys, thermal firebricks, boron carbide
Plastic	Thermoplastics, including elastomers	Acrylonitrile butadiene styrene, acrylic, celluloid, cellulose acetate, cyclic olefin copolymer, ethylene-vinyl acetate, ethylene vinyl alcohol, fluoroplastics, liquid crystal polymer, polyoxymethylene, polyacrylates (acrylic), polyacrylonitrile, polyamide, polyamide-imide, polyaryletherketone, polybutadiene, polybutylene, polybutylene terephthalate polycaprolactone, polychlorotrifluoroethylene, polyethylene terephthalate, polycyclohexylene dimethylene terephthalate, polycarbonate, polyhydroxyalkanoates, polyketone, polyester, polyethylene, polyetheretherketone, polyetherketoneketone, polyetherimide, polyethersulfone, polysulfone, chlorinated polyethylene, polyimide, polylactic acid, polymethylpentene, polyphenylene oxide, polyphenylene sulfide, polyphthalamide, polypropylene, polystyrene, polysulfone, polytrimethylene terephthalate, polyurethane, polyvinyl acetate, polyvinyl chloride, polyvinylidene chloride, styrene-acrylonitrile, (thermoplastic elastomers) styrenic block copolymers, polyolefin blends, elastomeric alloys, thermoplastic polyurethanes, thermoplastic copolyester, thermoplastic polyamides
	Thermosets	Amino resins, polyester fiberglass systems (sheet molding compounds and bulk molding compounds), vulcanized rubber, Bakelite, Duroplast, urea-formaldehyde foam, melamine resin, epoxy resin, graphic-reinforced plastics, melamine, polyimides, silicon resins, epoxy resins, cyanate esters, polycyanurates, polyurethanes, phenolic resins
Metal	Primary metals	Iron, steel, copper, lead, aluminum, nickel, tin, zinc, gold
	Noble metals	Metals resistant to corrosion and oxidation: gold, tantalum, platinum, silver, and rhodium
	Alloys	Alloys of steel, stainless steel, cast iron, tool steel, and alloy steel; stainless steel created by adding silicon with chromium, nickel, and molybdenum
	Ferrous metals	Steel and pig iron (with a carbon content of 1.5–4%) and alloys of iron with metals such as stainless steel
	Nonferrous metals	Metals that are not magnetic and are more resistant to corrosion, including aluminum, copper, lead, zinc, and tin
Composites	Metal, ceramic, and polymeric matrixes	Primary stock (metal, ceramic, and polymeric–plastic matrixes) with other stock (in the form of particles, fibers, sheets, rods, or other materials) added to the matrix to improve machinability, durability, and strength

TABLE 2.9. The Five Basic Process Classifications (BPCs)

BPC	Process Examples
Forming	Extrusion, stamping, spinning, die casting, injection molding, thread rolling, resin transfer molding, bending, laminating, fiber drawing, slip casting, hot pressing
Separating	Grinding, sawing, drilling, milling, blanking, punching, routing, chemical etching, filter pressing, dicing
Joining	Ultrasonic welding, hot melt adhesives, blind riveting, fusion sealing spray metallizing, filament winding
Conditioning	Radio frequency dielectric heating, sintering, vapor deposition coating, ion implantation, calcination, irradiation, radiation processing, cryogenic conditioning
Finishing	Electrostatic spraying, dip coating, roller coating, flame polishing, glazing, staining, electroplating, sand blasting

processes that work well with a chosen material. This is where major mistakes can be made and significant wastes can be generated. As was the goal before when we classified materials to reduce many materials to a few, we now group hundreds of processes with common characteristics using what we call *basic process classifications* (BPCs), which are shown in Table 2.9.

We use these five BPCs in classifying and selecting profitable and compliant materials and processes. Most of the manufacturing processes used today by secondary manufacturing firms fit in one of these basic process classifications. Within each process class are dozens or hundreds of processes that can perform the process action. Deciding which one to select is determined in part by the material, design requirements, and the amount of waste that will be created. This is the essence of the process selection decision. (Tables 2.10–2.14 provide examples of manufacturing processes that are used by each MMF. Please remember that these are just examples. There are thousands of different manufacturing processes.) Let's take a look at the BPCs and discuss what is unique about each process classification.

Forming Processes

Forming processes are used to shape, stretch, twist, and bend materials. They are also used to compress material in a mold or form. In most cases, forming processes do not involve the removal of material or a change in volume. Their job is to change the shape of the industrial stock through the application of pressure or force. Most forming processes are unique to the type of material being formed. For example, the manufacturing process called blow molding can be used with softened plastics to make products like huge fuel tanks, but it is not useful with ceramics or metals because of the behavior of these materials. Neither ceramics nor metals can be kept in a semiliquid state during forming and still result in a quality product. With these materials molds are used to hold the material together during casting or firing. Molds are not necessary when working with plastic processes such as blow molding.

Table 2.10 shows forming processes used with the MMFs.

Separating Processes

Separating processes are those processes that are used to remove material or volume. Some separating processes produce chips or other forms of waste. Other separating processes do not produce

TABLE 2.10. Forming Processes Used with Major Material Families (MMFs)

MMF	Manufacturing Process	Manufacturing Process
Metal	Open-die and closed-die forging	Swaging
	Coining	Stamping
	Thread rolling	Electromagnetic forming
Plastic (including rubber)	Extrusion	Thermoforming
	Injection molding	Open molding
	Compression molding	Rotational molding
Ceramic	Dry pressing	Band or tape casting
	Hot isostatic pressing	Slip casting
	Jiggering	Fiber drawing
Wood	Lamination	Cold bending
	Wet bending	Plasticizing
	Bonding	Steaming
Composite	Liquid infiltration	Filament winding
	Coextrusion	Chemical vapor deposition
	Elastomeric molding	Thermal expansion
	Open and closed molding	

TABLE 2.11. Mechanical Chip-Producing Separating Processes Used with Major Material Families

MMF	Manufacturing Process	Manufacturing Process
Metal	Sawing	Perforating
	Punching	Broaching
Plastic	Foam cutting	Granulation
	Vibratory deflashing	Grinding
Ceramic	Attrition milling	Fluid energy milling
	Pugging	Sandblasting
Wood	Drilling	Joining
	Shaping	Boring
Composite	Diamond-wire sawing	Waterjet cutting
	Turning	Knurling

any chips or waste at all. There are three major types of separating processes:

· Mechanical chip-producing separating
· Mechanical non-chip-producing separating
· Nonmechanical separating

Mechanical chip-producing separating processes (refer to Table 2.11 for examples) use a wedge-shaped cutting force to remove material. Mechanical non-chip-producing separating processes deliver a wedge-shaped cutting force on the material but no chip is created and there is no loss of material (examples are shown in Table 2.12). Nonmechanical separating processes do not use force and may or may not generate chips (refer to Table 2.13 for examples).

Joining Processes

Joining processes are used to fasten or fabricate industrial stock, parts, subassemblies, components, or partially finished workpieces together. (Examples of joining processes are shown in Table 2.14.) There are three types of joining processes: processes that rely on adhesion; processes that rely on cohesion; and processes that rely

TABLE 2.12. Mechanical Non-Chip-Producing Separating Processes Used with Major Material Families

MMF	Manufacturing Process	Manufacturing Process
Metal	Shearing	Stamping
	Blanking	Slitting
Plastic	Continuous die cutting	Shearing
	Stamping	Laser cutting
Ceramic	Mix mulling	Spray drying
	Filter pressing	Tempering
Wood	Abrading	Burnishing
	Rolling	Wood burning
Composite	Embossing	Silicon wafer dicing
	Shearing	Hydroabrasive machining

TABLE 2.13. Nonmechanical Separating Processes Used with Major Material Families

MMF	Manufacturing Process	Manufacturing Process
Metal	Etching	Electrochemical machining
	Electrodischarge machining	Ultrasonic machining
Plastic	Polymerization	Crystallization
	Mixing	Thickening
Ceramic	Thermal separation	Firing
	Filtration	Freeze drying
Wood	Air drying	Kiln drying
	Burning	Soaking
Composite	Ultrasonic machining	Electrostatic separation
	Ultrasonic cutting	Corona discharge machining

on mechanical joining. Adhesive and cohesive joining processes create permanent connections and are not used in situations where the parts or product must be disassembled for servicing or repair. Mechanical joining processes permit disassembly.

Conditioning Processes

Conditioning processes are processes that produce changes in the mechanical properties or molecular structure of a material. This may be something that occurs throughout the entire workpiece or

TABLE 2.14. Joining Processes Used with Major Material Families

MMF	Manufacturing Process	Manufacturing Process
Metal	Mechanical joining: locking pellet fasteners, bolts and nuts, cotter pin	Cohesion: resistance welding
	Cohesion: soldering	Cohesion: spin welding
Plastic	Adhesion: hot gas welding	Adhesion: fusion sealing
	Mechanical joining: pop and blind rivet	Adhesion: laminating
Ceramic	Adhesion: gluing	Cohesion: firing
	Adhesion: glazing	Cohesion: ultrasonic joining
Wood	Adhesion: laminating	Cohesion: hot melt adhesive
	Mechanical joining: bolts, screws	Mechanical joining: mortise and tenon
Composite	Cohesion: sol–gel processing	Cohesion: vacuum forming
	Adhesion: laminating	Adhesion: transient liquid phase bonding

only on the surface of the product. Conditioning processes are used to improve the strength, hardness, wear, and fatigue resistance of products. Case hardening, tempering, radiation processing, and corona discharge treatment are examples.

Finishing Processes

Finishing processes are used to improve the appearance of the product, to clean a workpiece to accept another coating, and/or to protect the product in its operating environment. Spray metalizing, plating, painting, and electrostatic powder coating are examples of finishing processes.

You may want to stop for a moment now to think about one of your products, how it is made, and how the manufacturing processes can be classified according to the BPC concept. What processes create the most waste in your production system? Keep in mind that our goal for organizing processes in this way is to make it easier to select processes and materials that result in a product that is profitable and minimize sources of waste. For additional information, www.matweb.com provides a searchable database of more than 59,000 metals, plastics, ceramics, composites, and wood materials.

DESIGN TEMPLATE FOR CLASSIFYING MANUFACTURING PROCESSES

We have covered the materials, the processes, and the classification system that we are using. Now we are ready to construct and apply the tool for carrying out an analysis: the "profitable and compliant process chart" (PCPC). Here in Chapter 2 we keep things simple to help you become familiar with the chart. In Chapter 7 we examine a more complex product to show how to improve the quality of analysis. At that point you will see how the selection of processes and materials can be used to evaluate alternatives and their waste potential for a particular manufacturing operation. It is the waste potential of a process or material that will determine compliancy with environmental regulations and the amount of resources required. This should provide you with a tool you can use to make decisions about production in your own environment and help your company become more environmentally compliant. If you don't already, you will soon recognize that regulations, while sometimes viewed as a nuisance, may really provide an open door to profitability.

To set the stage for our analysis we need some data from a manufacturing activity. Let us tell you about Sally's problem.

IT ALL BEGAN IN SALLY'S GARDEN

The need for a new product became critical when a gardener named Sally took an early morning walk through her garden. She was shocked to see that each of her prized tomato plants was drooping over and touching the ground! There was a definite need for a device to support her tomato plants. An image of a three-foot-long garden stake came to mind. She studied the plants and determined she would need 16 stakes in total. Sally realized that she could manufacture the stakes in her garage workshop. As she walked to the shop she mentally listed the requirements for the garden stake. She realized that she was doing exactly what the marketing department did at work.

Once she was in the garage she thought about the dimensions for the stake and what material she would use. She considered three stocks: hot-rolled angle iron, 3/4" ABS (acrylonitrile butadiene styrene) water pipe, and clear vertical grain fir. She had all three stocks available in her workshop. Sally decided that the stock for the garden stake should be wood, mostly because it is the traditional stock for garden stakes. With the dimensions in mind and the material selected, she had completed all of the initial product engineering tasks.

Next she ran through the processes she would use to manufacture the garden stake. The first was a separating process. She gathered her stock, three 1 × 6 boards, from the lumber rack mounted on the sidewall of the garage. One board was five feet long and the other two were six feet long. She determined that the boards were wide enough to get four strips out of each board; then she could cut the strips in half, giving her eight stakes per board. However, that meant that the five-foot board would only give her a stake that was 30" long. With this in mind she walked back out to the garden to see if a 30"-long stake would do the job. As she was walking she thought of all the times manufacturing called marketing and engineering to see if they could get a design deviation approved. After looking again at her 16 plants she decided the 30" stakes would work fine on all but four of the plants. These would need the longer stake. She smiled, thinking it would be wonderful if it was this easy to put through a "manufacturing change" at work.

Now it was time to start manufacturing stakes. The first process was turning the 1 × 6 boards into 1⅜" strips. She had a table saw that could easily do the job. She also had an axe that she could use to split the boards into strips, but she didn't give this any consideration before deciding to use the table saw. So she went ahead and moved the saw out of the garage and into the driveway. Sally smiled to herself thinking her workshop had more flexibility to set up to manufacture a new product than her employer's factory. With the table saw in position she ripped the boards lengthwise (a chip-producing separating process using a resource, electric power). This would be the first operation in the basic process classification—separating a material stock (wood).

Once the stock was separated into strips it was then crosscut to length. She had to use her old miter gage to do this, and rust prohibited the gage from turning. Sally remembered that her neighbor had borrowed her new miter gage for the table saw, so she had to use a handsaw to cut the point on the stakes (another separating process using a resource, human energy). The workpieces (work-in-process) were stored temporarily on top of the garbage can while she went back to the garage for a handsaw. When all the separating processes were finished it was time to clean up the sawdust on the ground. She used a leaf blower to blow this into the neighbor's rosebushes on the property line (waste recycled without using resources—gasoline and oil).

Part of Sally's design included painting the stakes green. This was intended to preserve the stakes so they could be reused next year and to make them more aesthetically pleasing. A trip to the local hardware store (using resources—gasoline and oil) was necessary since no green paint was available in the garage. She purchased a can of oil-based enamel (a source of VOCs) and a disposable brush. Then she painted (a finishing process) the stake and placed the can with the remaining paint in inventory in the garage. The brush was wrapped with newspapers and tossed into the garbage can. Later it was transported (using several resources) to the local dump.

THE ANALYSIS

Granted, the situation in this story is goofy and simplistic. However, it can be used to illustrate several of the major decision points

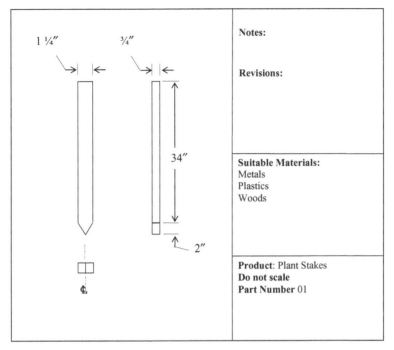

Figure 2.1. Sally's plan with available stocks.

in the design and manufacture of a product even though it is nothing more than stakes to support tomato plants. In this scenario the three key functions—marketing, product design, and manufacturing—were conducted by one individual. Although this is an ideal for communication and the exchange of information, many of the decision points were passed and the opportunities to be profitable and environmentally compliant were missed. Let's take a moment to study Sally's thinking when constructing her garden stakes. See Figure 2.1, Sally's plan for the stakes.

Wood was selected as the most appropriate stock for making the stakes. At this point Sally thought about the process flow that would be most effective given the processes and materials that were available. This began as a mental model and was later converted to note paper. Her plan included four operations, two material stocks, and six decision points. See Figure 2.2, for Sally's manufacturing process flow.

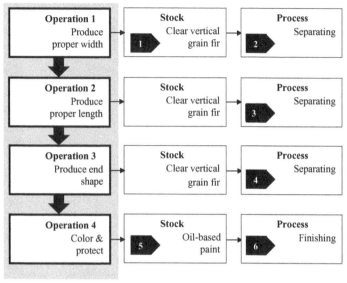

The manufacturing sequence consists of four processes (operations) that use two material stocks. This creates six decision points: ▶

Figure 2.2. Manufacturing process flow for Sally's garden stakes.

The decision points shown in Figure 2.2 are important in the design of a strategy to assess profitable and compliant manufacturing processes and materials. However, a methodology that would "mark" or provide value to these decision points is needed. The profitable and compliant manufacturing process (PCMP) analysis does this. This analysis is also the basis for instituting a collaborative approach to identify improvements. A PCMP analysis begins with the creation of an adapted version of the process flow chart, called the profitable and compliant process chart (PCPC). It is a useful tool for performing self-analysis on what is going on in production. The data captured in the chart can help individuals, task groups, and teams concentrate on waste reduction, cost reduction, and efficiency improvement.

Now we are ready to fill in the chart (Table 2.15) using data from Sally's story.

Table 2.15 shows data from each operation involved in making the product and classifies the wastes generated. The first column

TABLE 2.15. Sally's Profitable and Compliant Process Chart

Task	BPC	MMF	MP	Waste Potential Classification			
				There Are Materials or Drops not Recycled (Part Has Some Trimming or Material Removal)	Oils, Lubricants, and/ or Solvents Are Used	Energy	Pollution
1	Separating	Wood	Ripping using a table saw	Yes	No	Yes	Slight
2	Separating	Wood	Crosscutting using a table saw	Yes	No	Yes	Slight
3	Separating	Wood	Crosscutting using a handsaw	Yes	No	No	Slight
4	Finishing	Paint	Painting by hand using a disposable brush	Yes	Yes	Yes	Slight

BPC, basic process classification; MMF, major material family; MP, manufacturing process.

shows each operation as a numbered task. The chart lists the complete sequence of operations.

Column 2 provides information on the BPCs. Sally used three chip-generating separating processes and one finishing process. Column 3 indicates that wood was the MMF involved. Information provided in column 4 shows that three sawing processes were involved. Two of these were accomplished with a table saw and the third with a handsaw. A final finishing process involving painting was also used. The operations used two stocks: wood and oil-based paint. The disposable brush is an ancillary stock or *secondary material*—something that is not part of the product but is necessary and is used up in producing the product.

This is a very simple illustration of what might happen in the process of creating a manufactured product. However, much information can be gleaned from this simple illustration. One important concern that we have not discussed is the data associated with *waste potential classification*. Columns 5–8 contain information on the impact of production under the heading of "Waste Potential Classifications." Let's use row 1 to illustrate how this works. The BPC is "separating" and wood is the major material involved. The manufacturing process is "ripping" and the machine used to do this is the circular saw. Looking under column 5 we see that judgments were made about the materials and *material drops* (materials that were removed in the ripping process). We know that sawdust in the neighbor's rosebushes is an example of material drops not recycled (except as mulch). That is why we provided a "Yes" in this column. Looking across to column 6, we see that oils, lubricants, and/or solvents were not in this operation so we provided a "No" in this column on the chart. We know that electrical energy was required to run the saw. A "Yes" was added to the chart for this column. Looking on to column 8 we need to make a judgment about pollution resulting from task 1. On the ground is a pile of sawdust, from this process. We provided a rating of "Slight" in column 8.

The data portrayed with this type of chart could be used by task groups and teams concentrating on waste reduction. But the chart has some limitations when used to capture data from Sally's manufacturing operation. A major drawback is the subjective assignment of values in each column. Using a scale of "No, Slight, or Yes" is an example of rank-ordering data using an ordinal scale.

The *Stevens power law* states that there are four different types of scales that can be used to define how things can be measured or arranged (Stevens, 1957). Stevens called these scales nominal, ordinal, interval, and ratio. *Nominal* is a categorical scale that gives names to things, for example, the names of the 50 states of the United States. *Ordinal* is a ranking scale like 1st, 2nd, or 3rd place. *Interval* is a scale for measurable data, for example, temperature in degrees Celsius. The *ratio* scale is used to measure length, mass, energy, and so on. All statistical measures can be used for data that is represented on the ratio scale.

Returning to Table 2.15 we see that it contains nominal and ordinal data. Simple judgments about waste potential with responses such as "Yes" or "No" have limited value. What is needed is to upgrade the tool to interval (scaled) or ratio (rank-ordered) data. More details are also needed on levels of waste and waste streams. All of these factors will be addressed in the implementation model of the PCPC provided in Chapter 7.

NEXT STEPS

Other considerations must also be addressed in the final PCPC. First we need to spend more time understanding how a company must meet a variety of community, state, and federal regulations. Consequently Chapter 3 provides an overview of these regulations. These regulations should be viewed as adding to or possibly enhancing our definition of waste. Therefore the effort an organization puts into minimizing waste also helps it become environmentally compliant. Along with the summary of regulations Chapter 3 also looks at voluntary certification standards. These standards play a significant role in helping a company structure itself to meet its goal for waste minimization. Again, keep in mind that complying with environmental regulations is part of the practice of reducing waste.

SELECTED BIBLIOGRAPHY

Fourastié, J. (1949). *Le Grand Espoir du XXe Siècle*. Paris: Presses Universitaires de France. Reprinted as *Moderne Techniek en Economische Ontwikkeling* (1965). Amsterdam: Het Spectrum.

Gartman, D. (2009). *Market Folly*, April 16, 2009.

National Association of Manufacturers (NAM) (2011). Facts about manufacturing. Available at http://www.nam.org/Statistics-And-Data/Facts-About-Manufacturing/Landing.aspx (accessed December 21, 2011).

North American Industrial Classification System (NAICS) 2010. Manufacturing sectors 31–33. Washington, D.C.: Office of Management and Budget. Available at http://www.census.gov/econ/census02/naics/sector31/31-33.htm (accessed March 10, 2010).

Stevens, S. S. (1957). On the psychophysical law. *Psychological Review* 64(3):153–181.

USGS (2011a). *U.S. Geological Survey and Conference Board Report*, July 22, 2011.

USGS (2011b). *U.S. Geological Survey, Mineral Commodity Summaries*, January 2011.

CHAPTER 3

ENVIRONMENTAL REGULATIONS, STANDARDS, AND PROFITABILITY

INTRODUCTION

In the preceding chapter the discussion emphasized identifying and classifying resources and sources of waste in manufacturing. Now we want to examine how compliance with environmental regulations fits into a strategy for profitability. The question is, "can complying with regulations to protect our environment make a company profitable?" Well, we think the answer might be a qualified yes! It is yes if the methods and techniques the company uses to reduce or eliminate waste and the resources to manufacture a product are also applied to achieve compliance. Keep in mind that most regulations are intended to eliminate or reduce a pollutant or to curtail an activity that creates a pollutant. Change the word "pollutant" to "waste" and the efforts to achieve compliance by minimizing pollutants can help profitability.

Improving Profitability Through Green Manufacturing: Creating a Profitable and Environmentally Compliant Manufacturing Facility, First Edition.
David R. Hillis and J. Barry DuVall.
© 2012 John Wiley & Sons, Inc. Published 2012 by John Wiley & Sons, Inc.

The next question that might be asked relates to when and how a manufacturing company should factor in regulations. Shouldn't the product design and the specification of materials come first? The answer to the second question is, "not really." First of all, product designers routinely consider safety regulations at the conception of the product and this is also the time to consider environmental compliance. The improvement in profitability and compliance begins with recognizing the influence regulations have on the design of the product. That is why the profitable and compliant process chart (PCPC) was introduced in the previous chapter and is expanded later in this book. Compliance requires the designers of both the product and the manufacturing methods to work as a unit. In the example in Chapter 2 we had one person doing these tasks. This is the ideal that seldom ever happens. Nevertheless the selection of materials and processes needs to be an interdependent endeavor. This concept is not new. Years ago one of the popular "management fads" was *early manufacturing involvement* (EMI). IBM was a major proponent of EMI. They felt it reduced the time needed to bring a new product from the point of conception to shipment to the customer. But it did more than that; having product designers and manufacturing engineers and technologists working together improved the quality of the product and reduced the cost to manufacture it, as well as shortening the development time. Certainly there are companies that have adopted this concept and have retained it as part of their normal design process. Unfortunately, as with most fads, companies moved on to the next management craze without internalizing EMI in their product design. EMI is a necessary concept for designing products to comply effectively with the all the regulations.

The implication therefore is that both the product designers and the personnel in manufacturing functions in an organization need to understand and follow the rules, regulations, and good practices for their company to be environmentally compliant. This can be done profitably if the company is organized effectively to do this and the people responsible are committed to eliminating waste and minimizing resources. With this in mind our first step is to examine how a company can organize itself to accomplish this objective.

ORGANIZING TO COMPLY—THE MANAGEMENT FOUNDATION

As we have mentioned before, design requirements for a product generally start from a customer's need and are further shaped by market competition. However, there is another starting point. Innovation and creativity can also lead to development of new products that customers don't realize they want. These products can be truly novel and usually don't have any market competition. In either circumstance the choice of materials and processes that will be used to produce the design will be impacted by federal and state environmental and safety regulations. In fact certain materials and processes will be prohibited or severely limited by these regulations. Therefore designers need to have these regulations in mind when they consider the various material options for the product.

The geographic location of the plant that will manufacture the product can also influence the material selection. This is caused by the variability of state regulations and population density. Regional regulations often can be more limiting than federal. Therefore developing a general engineering design guide or a set of engineering standards based on government regulations is not simple.

However, the federal government in an effort to be proactive has identified what's called *maximum achievable control technology*, also known by the acronym MACT. These technologies provide a means to use materials that have been listed as "toxic" or "polluting." Therefore a listed material can still be used by employing a maximum achievable control technology. Selecting one of these materials will also define the processes and to a limited extent the geographical location where they can be used.

With this in mind it is possible to say that product design, facility design, and plant location are being tied together due to the regulatory requirements that are being imposed on companies. This can still be a positive development as we will see later.

The regulatory requirements that designers, manufacturing engineers, and technologists must be familiar with are in three main categories: clean air; clean water; and solid or hazardous waste management. This body of regulations requires a blizzard of reporting and record keeping requirements that each facility or company must maintain. The theory for the record keeping is analogous to the underlying concepts of process control. The facility or company must monitor their operations as mandated and document that they are in compliance so that there is proof that the regulations are being met. If they have an incident or an event where they are not in compliance, they must document how they took action, must make the necessary notifications, and must bring the process or operation back into compliance. This is why it is similar to the concept of process control, which is based on prevention as opposed to inspection. It is also part of the underlying approach found in *total quality management* (TQM) and can be used as a guide in creating a management system to be profitable and compliant.

Before the introduction of TQM, quality systems were stand-alone organizational structures based on inspection. The premise was to create a series of checkpoints to catch defects and isolate them from good product. The quality and inspection effort was an overhead cost that did not add value to the product. In fact in most cases companies saw a certain level of rejects as a given in the cost of goods sold. The wages paid for inspection were considered necessary and appropriate for a quality-conscious organization. As recently as 25 years ago companies actually bragged about how many people they employed as inspectors. The TQM approach, however, focused on creating processes and operations that were capable of making the product as designed. Creating and maintaining capable processes, detecting problems before they caused defects, and problem solving are an integral part of the total quality approach. The role of inspection was reduced to a practice of confirming that the quality management system was working.

The important aspect of this approach was the change in the role of management. They did not accept rejected product as a given and recognized that inspection is an overhead cost—actually a form of waste. Managers slowly began to realize the magnitude of waste that existed in manufacturing. The tools they used to find this waste and eliminate it are addressed in more detail in later chapters. However, the companies that adopted total quality systems found that in addition to better quality and reduced overhead expense there was another benefit. Process control provides records that a company can use to demonstrate the quality of its operations to a customer. Being able to document quality performance became an effective selling point for manufacturers. Think of the parallels for a company that adopts an environmental management system that is based on prevention not inspection.

FORMALIZING THE MANAGEMENT APPROACH—THE ISO STANDARDS

The next step for many companies committed to quality management was to seek certification from the International Organization for Standardization (ISO), which is headquartered in Geneva, Switzerland. The ISO 9000 series of standards that became so popular in the 1990s became the definition of quality-oriented management systems. Companies can work toward being certified under one or more of the standards. The value for achieving certification implies but doesn't guarantee that a company has in place a management system that can be effective in reducing waste and providing a quality product to its customers.

The ISO standard's purpose is to form an error-free manufacturing environment based on a process of continuous improvement. The standards emphasize the need to document the company's procedures and the ability to demonstrate that these procedures are being followed. Briefly the ISO 9001:2008 standard provides a framework for establishing a disciplined approach for managing a manufacturing operation so the products produced will meet customers' expectations. For well-managed companies meeting the ISO standard is an auditing process not a burdensome addition to their management systems.

ISO 14000 Series of Standards

The philosophy that is inherent in the ISO quality standard is similar to what is required in an environmental management system (EMS). If a company is already certified under ISO 9001, the work to incorporate an EMS will not be overly burdensome. They will have many of the management procedures in place to establish the EMS found in the ISO 14001-2004 standard that is aimed at reducing the pollution and waste a business generates. What remains to be accomplished is the planning and implementation of environmental objectives and targets. The creation of an EMS requires the company to change its focus. The ISO 9000 standard is product oriented while the 14000 series is aimed at how a product is produced.

The first standard in the 14000 series is ISO 14001-2004, which specifies a process for controlling and improving a company's environmental performance. The following are major sections of the standard:

- General requirements
- Environmental policy
- Planning
- Implementation and operation
- Checking and corrective action
- Management review

The first step a company should take would be to purchase the actual standard from the ISO. With the standard in hand the work can begin; however, most companies find that some additional explanation and guidance are needed to meet the standard's requirements. There are businesses and consultants available that offer this help. They can provide training and education and can assist in setting an EMS and procedures to meet the standard's requirements. Once a plant has adopted and implemented all of the standard's requirements it is ready for its audit by an accredited auditor. Certification auditors must be accredited by the International Registrar of Certification Auditors. Once a company has successfully implemented the standard, the following are some of the benefits that can be realized:

- Reduced raw material and resource use
- Reduced energy consumption
- Improved process efficiency
- Reduced waste generation and disposal costs

Generally ISO 14001 is the standard that most manufacturers would select to follow. However, if you did a web search you would find a long list of standards in the series. Some of these are primarily informational but the following three would be of interest.

- ISO 14001 was first published in 1996 and revised in 2004. It specifies the actual requirements for an environmental management system. It applies to those environmental aspects which an organization has control over and can change.
- ISO 14004, also published in 1996, provides direction on the development and implementation of environmental management systems and co-ordination with other management systems.
- ISO 19011:2011 offers guidelines for quality and/or environmental management systems auditing. It supersedes a number of standards, including ISO 14010, 14011, and 14012.

OVERVIEW OF MAJOR ENVIRONMENTAL REGULATIONS

In the first half of the twentieth century air pollution was generally considered a local urban problem. However, World War II changed the climate in southern California. Industry and population growth due to the war effort created "smog," a mixture of smoke and fog, which became part of Los Angeles' identity. In some of the worst instances the smog limited visibility to less than half a mile. To control this pollution California passed the first state air pollution law in 1947. The law made municipal governments responsible for the passage and enforcement of pollution control legislation.

The first effort of the federal government to control air pollution started with the passage of the Air Pollution Control Act of 1955. This authorized federal research programs to investigate the health effects of air pollution and provide technical assistance to state governments. The Clean Air Act (CAA) of 1963 replaced the Air Pollution Control Act of 1955. This legislation required the Secretary of Health, Education, and Welfare to describe air quality criteria on the basis of scientific studies.

In 1965 the Motor Vehicle Air Pollution Control Act was enacted. This was the beginning of the federal government's efforts to reduce automobile emissions. The Federal Air Quality Act was put into effect in 1967 to establish a means for identifying "air quality control regions" based on meteorological and topographical factors.

President Richard Nixon proposed the creation of the United States Environmental Protection Agency (EPA). Through the use of an executive order the agency came into being in early December 1970. Now the EPA is involved in environmental assessment, research, and enforcement of environmental laws. The agency accomplishes its mission in consultation with state, tribal, and local governments.

Today there is a fairly comprehensive approach to environmental protection that impacts virtually all aspects of business and industrial activity. The next sections provide a very general overview of the federal acts that regulate the pollutants and toxic chemicals in air, water, and solid waste generated by manufacturers and industries. More detailed information can be found on the Federal and state EPA websites. Trade and professional organizations are also possible sources of industry-focused information.

These organizations are also a good starting point to learn about maximum achievable control technologies (MACT) that would apply to a particular industry.

Clean Air Act Overview

The Clean Air Act (CAA) consists of six sections, known as Titles. These Titles direct EPA to establish national standards for ambient air quality and for the EPA and the states to implement, maintain, and enforce these standards. A very brief explanation of what each title of the Clean Air Act includes the following:

- Title I establishes National Ambient Air Quality Standards (NAAQSs) that set limit levels for six "criteria pollutants." They are particle pollution (also called particulate matter), ground-level ozone, lead, carbon monoxide, sulfur oxides, and nitrogen oxides.
- Title II applies to mobile sources, such as cars, trucks, buses, and planes.
- Title III directs the EPA to create a list of sources that emit any of 188 hazardous air pollutants (HAPs). Note that the original list contained 189 HAPs; however, one item was dropped from the list so the total became 188. Examples of hazardous air pollutants include benzene, which is found in gasoline; toluene, which is used as paint thinner; and methylene chloride, a solvent and paint stripper.
- Title IV is designed to reduce the formation of acid rain by requiring power plants and other utilities to reduce sulfur dioxide emissions.
- Title V created a national, operating permit program for all major sources of emissions regulated under the Clean Air Act. EPA defines a major source as "a facility that emits, or has the Potential To Emit (PTE) any criteria pollutant or hazardous air pollutant (HAP) at levels equal to or greater than the Major Source Thresholds (MST)." These permitting

requirements are also based on the facility's location as well as the type of pollutant that is emitted. The categories and criteria can be found at the EPA website (see http://www.epa.gov/oaqps001/permits/obtain.html). It is also necessary to check the state's Title V EPA permitting requirements.

- Title VI is intended to protect stratospheric ozone by phasing out the manufacture of ozone-depleting chemicals and restricting their use and distribution. Title VI requires EPA to list all regulated substances along with their ozone-depletion potential, atmospheric lifetimes, and global warming potentials.

The Clean Air Act requires each state to create a state implementation plan (SIP) to identify sources of air pollution and determine what reductions are required to meet federal air quality standards. Title I also authorizes EPA to establish "new source performance standards" (NSPSs), which are uniform emission standards for new stationary sources within particular industrial categories. New source performance standards are based on the pollution control technology available to that category of industrial source.

As mentioned in Title III of the Clean Air Act the EPA has created a list of sources that can emit any of 188 HAPs that have been identified. The EPA is also required to develop regulations for these categories of sources. By 2010 the EPA listed 174 categories and developed a schedule for the establishment of emission standards. The emission standards are developed for both new and existing sources based on the MACT. MACT, as mentioned, is defined as the control technology that can achieve the maximum degree of reduction in the emission of HAPs. The cost of the control technology is a factor in selecting a particular technology (see http://www.epa.gov/apti/course422/apc4e.html). Facilities that fall into one of the identified categories and emit a hazardous air pollutant in more than the threshold quantity will have to prepare a risk-management plan for each air pollutant used at the facility.

Clean Water Act Overview

The Clean Water Act (CWA) was put into law to restore and maintain the physical and biological integrity of the nation's surface waters so that these waters would be healthy enough for fishing or swimming. The pollutants and materials or chemicals regulated under the act include the following:

- Priority pollutants, which include 126 listed chemicals (see http://water.epa.gov/scitech/methods/cwa/pollutants.cfm)
- Various toxic pollutants
- Conventional pollutants, biochemical oxygen demand (BOD), defined as the amount of oxygen required by oxygen-dependent microorganisms to decompose the organic matter in polluted water or sewage
- Total suspended solids (TSS)
- Fecal coliform
- Oil and grease
- Water-quality criterion for pH
- Nonconventional pollutants, which includes all pollutants that are not included in the list of conventional or toxic pollutants

The Clean Water Act also regulates storm-water runoff coming from either direct or indirect discharges that find their way into waters protected by the act. The pipes or man-made ditches that a factory uses to carry away this water are considered to be "discrete point sources or conveyances." Therefore the facility is required to comply with the National Pollutant Discharge Elimination System (NPDES) by controlling runoff to qualify for a permit. The NPDES permits are issued by either the EPA or an authorized state. The EPA has presently authorized 40 states to administer the NPDES program. These programs contain industry-specific technology and water-quality limits.

A facility that intends to discharge into the nation's waters must obtain a permit prior to initiating its discharge. A permit applicant must provide quantitative analytical data identifying the types of pollutants present in the facility's effluent. The permit will then define the conditions and effluent limitations under which a facility must operate. Water-quality criteria and standards vary from state to state, and by location within a state depending on the classification of the body of water receiving the discharge. Most states follow EPA guidelines for aquatic life and human health criteria based on the 126 priority pollutants.

The Clean Water Act also regulates the discharge into a publicly owned treatment works. The national pretreatment program requires industrial users to meet certain pretreatment standards before wastewater can be discharged into a sewer system. The goal of the pretreatment program is to:

- Protect municipal wastewater treatment plants from damage that may occur if hazardous, toxic, or other wastes are discharged into a sewer system
- Protect the quality of sludge generated by sewage treatment plants

A plant's discharge to a publicly owned treatment works are regulated primarily by the public treatment facility itself rather than by the state or federal EPA. However, the EPA has technology-based standards for industrial users of publicly owned treatment plants. Different standards apply to existing and new discharge sources within each industrial category. Companies should be sure to determine if there are more-restrictive "local limits" to the EPA pretreatment standard. These restrictions would be in place to help the publicly owned treatment plant to meet the effluent limitations in their NPDES permit. If the state is authorized to implement either the NPDES or a pretreatment program it may impose requirements that are more stringent than federal standards. For this reason it is possible that a plant may be an environmentally "friendly" facility one year and find that it does not meet the local standard the next year. Therefore it is the company's responsibility to keep informed and monitor its water usage and sewage discharge as well as what runs off its roof and parking lot.

Solid and Hazardous Waste Management Overview

Hazardous waste is discarded materials that are dangerous or potentially harmful to our health or the environment. Hazardous wastes can be liquids, solids, gases, or sludge. They can be discarded commercial products, such as cleaning fluids or pesticides, or the by-products of manufacturing processes. Also, it is important to point out that all states have solid or hazardous waste regulations that in many cases are different or more stringent than the federal rules.

In general hazardous waste is defined under the federal law as a solid or liquid waste exhibiting one or more of the following characteristics: being flammable, corrosive, reactive, and/or toxic. Waste oil and solvents are included in this definition of hazardous waste. A list of links that describe the various forms of waste can be found at the EPA website on wastes and hazardous waste (see http://www.epa.gov/osw/hazard/index.htm).

The following are some of the terms and phrases that are frequently used in the hazardous waste regulations:

- Resource Conservation and Recovery Act (RCRA) was enacted in 1976 and is the main federal law governing the disposal of solid and hazardous waste.
- Hazardous wastes are divided into listed wastes, characteristic wastes, universal wastes, and mixed wastes. Specific procedures have been established to determine how a waste is identified, classified, listed, and de-listed.
- Hazardous waste generators are divided into categories based on the amount of waste they produce each month. Different regulations apply to each generator category.
- Hazardous waste transporters move waste from one site to another by highway, rail, water, or air. Federal and state (if

applicable) regulations govern hazardous waste transportation, including the waste manifest system.
- Treatment, storage, and disposal (TSD) requirements govern the treatment, storage, and disposal of hazardous waste at facilities established and permitted for this purpose.

Solid waste can be disposed of in several ways, but in all cases the disposal method must make the waste less of an environmental threat. One common method of treating hazardous waste is combustion or incineration, which is used to destroy hazardous organic material and reduce the waste's volume. Another common method is placing the waste in a landfill. Landfill facilities are usually designed to permanently contain the waste and prevent the release of harmful pollutants to the environment. The EPA has specific requirements and restrictions concerning land disposal (see http://www.epa.gov/wastes/hazard/tsd/ldr/index.htm).

This overview of the three major areas of regulations is not at all detailed or intended to provide specifics on how an industry can comply with these regulations. The objective is to provide a sense of the scope of these regulations. A review of the federal and state websites for a particular act or regulation can provide more specifics on how regulations will impact an industry or a particular plant. Also a web search will show a variety of resources that are available to help a company bring a specific manufacturing process into compliance with a current regulation. Also using the phase "plain English" in any web search for government regulations can help.

SUMMARY—COMPLIANCE CAN MEAN PROFITS

In the first chapter there was a catalog of sorts listing the various sources of waste. Waste has come to be another word for cost, particularly since disposal of waste is never free. This chapter has outlined the regulations concerning waste streams leaving the plant site that are gaseous, liquid, or solid. In some cases there are limits or "caps" to particular categories. Even clean water that goes down a sanitary sewer or flows into a storm drain can or will be regulated by local and/or federal law.

Knowing the scope of the regulations and what is regulated or in some instances prohibited expands the list of what should be considered waste and its potential cost to the manufacturing operation. However, controlling these waste streams to meet regulatory requirements does not necessarily reduce costs. In fact if the control is achieved using abatement technology to meet the regulation, costs will increase. This is why companies fight new environmental regulations.

To make this a profitable approach, the waste stream has to be either reduced to an allowed limit or eliminated. Elimination gets rid of the cost of "buying" the waste since the vast majority of waste comes from the materials or resources the company has purchased. Getting rid of the waste means less cost and furthermore the company doesn't have to purchase the equipment, do the maintenance, or buy the energy needed for the abatement technology.

In summary, knowing the regulations means you are able to identify more waste items that can be eliminated or reduced. Or, said another way, it provides more opportunities to reduce costs and increase profits. It is not that regulations make profits—it's the process of planning, carrying out the plan, documenting performance, and taking prompt corrective action to reduce compliance costs and waste that creates profits. The essential concept is recognizing what constitutes waste and then working continuously to prevent it that makes manufacturing operations profitable. This approach is the basis for good manufacturing management. It's the way to do business. So if a company has this approach built into its everyday operations, then complying with environmental and safety regulations will not be an overhead cost or an add-on management system.

SELECTED BIBLIOGRAPHY

International Organization for Standardization. ISO 14000 essentials. Available at http://www.iso.org/iso/iso_14000_essentials (accessed January 8, 2012).

International Organization for Standardization. ISO 9000 essentials. Available at http://www.iso.org/iso/iso_9000_essentials (accessed January 8, 2012).

U.S. Environmental Protection Agency. HAPs list. Available at http://www.epa.gov/apti/course422/apc4e.html (accessed January 8, 2012).

U.S. Environmental Protection Agency. History of pollution control. Available at http://www.epa.gov/apti/course422/apc1.html (accessed January 8, 2012).

U.S. Environmental Protection Agency. History of the EPA. Available at http://www.epa.gov/aboutepa/history/origins.html (accessed January 8, 2012).

U.S. Environmental Protection Agency. Land disposal restrictions. Available at http://www.epa.gov/wastes/hazard/tsd/ldr/index.htm (accessed January 8, 2012).

U.S. Environmental Protection Agency. MACT. Available at http://www.epa.gov/ttn/atw/mactfnlalph.html (accessed January 8, 2012).

U.S. Environmental Protection Agency. Priority pollutants. Available at http://water.epa.gov/scitech/methods/cwa/pollutants.cfm (accessed January 8, 2012).

U.S. Environmental Protection Agency. Solid and hazardous waste. Available at http://www.epa.gov/osw/hazard/index.htm (accessed January 8, 2012).

U.S. Environmental Protection Agency. Title V permit. Available at http://www.epa.gov/oaqps001/permits/obtain.html (accessed January 8, 2012).

CHAPTER 4

CASE STUDIES

INTRODUCTION

Chapter 2 began with a discussion on what constitutes manufacturing. The initial purpose was to acknowledge what most people in manufacturing already know—manufacturing is a complex and challenging business. The North American Industry Classification System certainly illustrates this complexity; adding to this all the attendant materials and processes, even the most competent manufacturing engineer or technologist will be challenged to recognize all the opportunities for limiting waste. The *profitable and compliant process chart* (PCPC) offers a way to find these opportunities. As you saw in Chapter 2 it provides a systems approach that models manufacturing based on five basic manufacturing process classifications coupled with a simple classification structure or grouping of material stocks. The chief benefit of this approach is its ability to highlight opportunities for designers and

Improving Profitability Through Green Manufacturing: Creating a Profitable and Environmentally Compliant Manufacturing Facility, First Edition.
David R. Hillis and J. Barry DuVall.
© 2012 John Wiley & Sons, Inc. Published 2012 by John Wiley & Sons, Inc.

manufacturing technologists to minimize waste and resource use. It is particularly applicable for manufacturers of hard good consumer products (Stage 2 manufacturing).

These opportunities are actually decision points that appear each time a new material stock or process is indicated in the sequence of operations. Each decision point requires manufacturers to examine alternatives that have the potential to reduce waste. For an analysis of alternatives to be worthwhile, the potential for generating waste has to be quantified for each alternative. This task would have been overwhelming just two decades ago but today's data collection and computer systems make acquisition and handling needed data almost routine or at least doable at a reasonable expense.

You may be wondering if the PCPC is universally applicable to the other two stages of manufacturing. The first stage as you recall is defined by the very material it processes to make a stock. An example is the steel industry. In this industry the raw materials are defined to create a specific stock, steel. In the last stage of manufacturing, Stage 3, the product limits or defines the processes. In the automobile industry a Stage 3 business would be an automobile dealer. In each of these two stages there are specific constraints to either the materials or processes which limit the number of decision points. This is why the PCPC has limitations for Stage 1 and Stage 3 manufacturers.

There is a further concern. Look at Stage 2 in Figure 4.1, which summarizes the manufacturing sequence. You will notice that there are two groups in Stage 2. The first group (Group 1) consists of manufacturers that design and build their own products, which the PCPC approach addresses directly. These companies have all the decision points available that allow them the opportunity to make effective choices of material stocks and processes.

The second group (Group 2), however, makes parts and components or in some situations the complete product for another company that has designed and developed the product.

What is important to understand is that the number of decision points decreases for the second group unless the Group 2 company

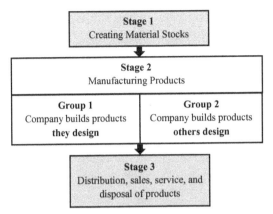

Figure 4.1. Manufacturing sequence showing the two Stage 2 manufacturing groups.

is able to collaborate with the company designing the product. This would then make the PCPC an equally effective tool for the Group 2 manufacturers. There is one other circumstance to be considered. A Group 2 company is normally considered to be a *manufacturing specialist*. For example, a manufacturing specialist could be a plastics molding plant. A plant of this type would have one primary process, a range of injection molding machines that produce molded plastic parts from a range of primary stocks, thermoset or thermoplastic materials. The selection of this manufacturing specialist would be the result of a decision made by the product designers. In this situation the designers of the product would have two decision points: first, the selection of a stock; second, the selection of a process.

When the specialist company uses the PCPC it will revert to its fundamental form, a process flow chart. The flow chart is useful but it is not a powerful approach for waste minimization. Some other means will have to be brought into use. The following case studies introduce and illustrate several approaches that are used by companies in all stages of manufacturing to be profitable and environmentally compliant. The case studies that follow serve to introduce the concepts and techniques and illustrate how they are used. In each of the following cases the company or organization being described is pursuing the same goal—minimizing waste and resources.

The examples given in the following sections include a Stage 1 manufacturer, three Stage 2 manufacturers, and a special case involving Stage 3. Each case demonstrates several methods an organization can use to minimize waste. As you read through them, look for some commonalities. There seems to be a theme or a pattern to the approach taken by these organizations.

CASE STUDY 1

The first case is actually an overview of a case taken from the *Guidebook on Waste Minimization for Industries* published by Singapore's National Environment Agency (NEA) in 2003. The guidebook contains several cases that illustrate a variety of practices that companies can use to reduce waste. The guidebook also develops a compelling argument that waste minimization is a money-saving endeavor for a company. One case in particular, Chevron Oronite Pte Ltd, is a good example of the foundational approach a company can use to be environmentally responsible (see http://app2.nea.gov.sg/data/cmsresource/20100901491845260649. pdf [NEA, 2010]). Chevron Oronite has a plant located on Jurong Island, which is connected by a causeway to the mainland of Singapore. The facility makes additives and material stocks for petroleum products. Consequently Chevron's facility is an example of a Stage 1 manufacturer.

Introduction

The Chevron Oronite Pte Ltd Singapore plant is a fully integrated manufacturing facility producing a comprehensive range of additives for the petroleum industry. The company has a corporate environmental policy that emphasizes the importance of conserving natural resources. A fundamental part of the company's efforts to implement this policy is the adoption of a formalized approach to environmental management. Their management system received ISO 14001 certification. Briefly an ISO-certified management system is intended to control the environmental impact of a plant's activities while continuously improving its environmental performance. A characteristic of this form of management is its systematic approach to establishing environmental objectives.

A second foundational aspect of the company's policy of waste minimization is their belief "that waste minimization evolves from the initial project design stage" (see NEA, 2010, p. 1). The company's project team "adopted the relevant engineering standards and sound environmental practices" in the design and construction of the plant. Environmental compliance and the minimization of waste begin with design regardless of the stage of manufacturing. For Stage 1 manufacturers the primary design focus will be on its facility as opposed to product design for Stage 2 manufacturers.

The Chevron plant was designed to reduce waste at the source. In practice the company monitors the waste products it generates and then follows a program of continuous improvement toward a goal of zero discharge. They indicate that this can be accomplished by designing processes to eliminate the creation of waste or by changing the process so that the waste that is generated can be recycled. However, the company anticipated that these waste objectives would not be met immediately so they constructed an on-site incinerator.

A third facet to the company's approach to waste minimization and environmental compliance is its commitment to using best practices. It is a signatory to the Responsible Care Codes of Practice. The Responsible Care Codes of Practice is a chemical industry initiative to continuously guide and improve a facility's day-to-day operations in terms of health, safety, and compliance with environmental standards.

Responsible Care was first conceived in Canada in 1985 to address public concerns about the manufacture, distribution, and use of chemicals. In 1988, the United States' chemical industry implemented its own initiative for responsible care to achieve improvements in environmental, health, and safety performance that go beyond the levels of compliance required by the U.S. government. Currently the Responsible Care Codes of Practice have been adopted by over 50 national chemical manufacturing associations and through their membership these practices are now in use by thousands of facilities around the world (see http://www.icca-chem.org/en/Home/Responsible-care/).

The American Chemistry Council (ACC; Chevron Oronite Company LLC is a member of the ACC) has also established a set of Principles of Responsible Care that their member companies can elect to follow. These principles are intended to improve a company's environmental, health, safety, and security performance (see http://responsiblecare.americanchemistry.com/Responsible-Care-Program-Elements/Guiding-Principles/default.aspx). The following are a few of the key elements taken from these principles:

- Design

  Create products that can be manufactured, transported, used, and disposed or recycled safely.

  Create and operate facilities in a safe, secure, and environmentally sound manner.

- Distribution and use of material stocks

  Work with customers, carriers, suppliers, distributors, and contractors to foster the safe and secure use, transport, disposal, and provide access to hazard or risk information relating to the use of the material stock.

- Safety and the protection of the environment

  Develop a culture throughout all levels of the organizations to continually identify, reduce, and manage safety risks.

  Promote pollution prevention, minimization of waste, and the conservation of resources over the life cycle of the products produced.

Work with governments at all levels in developing effective and efficient safety, health, environmental, and security standards, laws, and regulations.

A unique aspect of the Responsible Care Management System that has evolved from these principles is the use of benchmarking against the best practices of leading private sector companies. The American Chemistry Council has also expanded the Responsible Care program to provide its members the opportunity to achieve third-party certification to both Responsible Care and the ISO 14001 standard. This means that a facility can develop and implement an environmental management system that meets the criteria for responsible care and the ISO 14001 standard. This standard/specification is RC14001. Third-party registrars, who are recognized as registrars to both ISO 14001 and RC14001, can perform the audit leading to certification.

There is always a question of how effective manufacturing facilities can be in terms of being good neighbors with the environment. The Jurong District (east and west)—and Jurong Island—is located in the southwestern part of Singapore. The Chevron Oronite plant is just one of many industrial facilities located in this area. Also in this area, just a couple of kilometers north of Jurong Island, is the Jurong Bird Park. When one walks through this beautiful lush park it provides some evidence that well-managed industrial facilities can be good neighbors to the environment.

Waste Minimization Programs

An environmental management system relies on the day-to-day commitment and practice of waste minimization by everyone in an organization. The Chevron Oronite plant uses employee-driven environmental management programs (EMPs) as a means to generate this commitment. One program is the Clean and Green Workplace Campaign (CGWC). This program is designed to help promote environmental awareness and an orderly workplace environment. The objectives of the Chevron Oronite waste minimization programs focus on employee ownership, behavior, use of good work practices, continuous improvement, and employee participation.

The CGWC has put a lot of emphasis on the adoption of good work practices and behavior. An example that was given describes the efforts to prevent spills and leaks in production and storage areas. The intent of this program appears to be based on prevention as opposed to policing incidents as they occur. This concept of creating and maintaining an orderly workplace is described in more detail in the next case study.

Reuse and Recycling Activities in the Office

Involving the employees in the office is also essential to instill the culture of waste minimization in the organization. Chevron believes that nothing is too small to reuse or recycle. For example, the company makes a point to reuse paper that is printed on one side only. In their offices, separate collection boxes are provided for paper to be reused and for paper to be recycled. In the Chevron case study they state "the value of recycling paper to protect the environment is fully understood by our employees and they are very committed to this program. With every 20 reams of paper reused, we save a tree or $96 ($4.80/ream). This makes good business sense on cost saving, and leads to a significant reduction in paper usage." (see NEA, 2010, p. 2).

Reduction and Reuse of Packaging

Chevron has used a team approach as part of their EMPs. These teams have found ways to reuse or recycle raw material packaging such as woven polypropylene bags, drums, and wooden pallets. In one example an EMP team found some local vendors that were willing to take pallets for reuse when they were given the pallets free-of-charge. One example cited that 17,000 pallets were reused instead of disposed of as solid waste. This resulted in $140,000 annual savings by these vendors.

Comment

Waste minimization begins with the design function in manufacturing. In this case the design centered on the facility as opposed
to the product. Therefore the key aspect of this case is the company's statement that the facility was designed to minimize waste.
For a Stage 1 manufacturer, facility design is paramount. Another
significant feature of this case study is the company's establishment of an environmental management system based on industry-
specific best practices.

CASE STUDY 2

The next case study describes an assembly plant, a Stage 2 manufacturer that assembles kitchen dishwashers. The plant is a unit of
a global manufacturer of kitchen appliances. This company designs
and builds its products so it is a Group 1 manufacturer, one that
selects both the materials and the processes that will be used to
make the dishwasher. This would imply that the company has
taken advantage of all the decision points in the PCPC. However,
on the local level the assembly plant operates as a Group 2 manufacturer since it has little control over the selection of the materials or processes it uses. Nevertheless there is still much that can
be done.

The analysis in this case was carried out by a graduate student
looking for examples of waste reduction. He had contacted the
company inquiring if he could visit the plant and see firsthand a
current effort to reduce waste. The following is his report.

Introduction

The Electrolux Group, a Swedish company, is a producer of
kitchen, fabric care, and cleaning appliances. In 2010 its sales
volume totaled nearly $16 billion and employed approximately
52,000 employees worldwide. In the United States the Electrolux

Home Products Group has a plant in Kinston, North Carolina, that builds Electrolux's line of dishwashers.

I visited the plant on June 16, 2011, and met with a staff Industrial Engineer. He showed me a study that was conducted on the "Main Assembly" section of the dishwasher assembly line to find ways to reduce waste. After reviewing the study we toured the line and spoke with several of the employees that participated in the study. They explained the purpose was to improve operations using the first three steps of the *5S process* as a way to eliminate waste.

The Study and Methods Used

The 5S process is a five-step approach developed by the Japanese for organizing a work area to improve its effectiveness. The Japanese name for each step begins with the letter "S" so not surprisingly it is called the 5S process. Therefore all the English equivalents also had to start with an "S":

- Seiri—a rough English translation is sorting or arranging and the term used is *Sort.*
- Seiton—the English translation is orderliness and the term used is *Straighten.*
- Seiso—the English translation is cleaning and the term is *Shine.*
- Seiketsu—the English translation is clean and the term is *Standardize.*
- Shitsuke—the English translation is discipline and the term is *Sustain.*

The 5S sequence therefore describes how to organize a work space to improve productivity by identifying and storing items in a specific location, maintaining the orderliness of the work area, and then preserving the new order. The people at Main Assembly began the study with a discussion about standardization, which created a clear understanding among the employees about how the work should be done. It also instilled ownership of the process in each employee.

Electrolux focused on the first three steps (Sort, Straighten, and Shine) of the process for improvement. They explained that the assembly area is a very busy place that accumulated a lot of scrap material from packaging, wrappers, and paperwork, along with boxes of extra materials. These extra items were not kept in any type of organized bins or boxes once they were released to production. This created a sense of clutter but according to one of the study participants, "it was just part of the job and we never thought much about it." The clutter of items was all part of the assembly process that had to be overcome to complete their work. There was not a standard location or designated location for any of the materials.

The team started working on the first "S" (Sorting) by "red tagging" items that needed to be stored together and then developed a plan to keep them organized. They used existing boxes and bins to sort and store these materials in a place that was centrally located for all workers. Next they identified items in the work area that needed to be replaced. They found several workbenches that were damaged or did not have enough storage space for the tasks to be completed. These benches were replaced with ones designed and built to provide the space needed.

Next they began the Straightening step by creating a plan to keep all like items together on storage racks. They posted lists so all employees would know what was included at a location. The team also developed a "Sweep & Clean" checklist, to cover the last "S" in their plan. The team thought it best to develop a checklist so that they could track completion and make employees accountable by signing off when the steps were completed. This was to insure that all the improvements made during the study would be continued.

The other part of the study was to eliminate waste. The team found in the first step (Sort) of the study that there were too many parts located at each work station and there was no formal control of these parts. My escort explained to me that previously when parts were released from the warehouse they were charged to that department not to a specific "build quantity." The various springs, clamps, and valves that were lost or did not make it into a dishwasher still counted as a cost item and reported on a monthly basis. The team determined that they only wanted the amount of parts they needed for a specific week's worth of work. This reduced the part count on the floor about 50%. As a result there was a 30% reduction in assembly part costs applied to the actual production. My guide explained that the stock room did not have a way to report on the excess material that was issued so this process actually "rightsized" the inventory on the production floor. The results are shown in Table 4.1, which was part of a presentation that my guide gave to the team after the project was completed.

The final portion of the study was the identification of specific job duties and tasks on the assembly line. This led to the development of written job descriptions. After the job descriptions were written the group conducted time studies to identify each member's cycle times on various steps of the assembly process. These

TABLE 4.1. Comparison of the Inventory Levels before and after Applying the 5S Sequence

EMS Continuous Improvement Workshop		
Main Assembly		
Component Name	Inventory Before	Inventory After
Springs		
Type 1	1,600	800
Type 2	2,400	1,200
Type 3	2,400	1,200
Clamps		
Type 1	3,000	1,500
Water valves		
Type 1	2,400	900
Heater shields		
Type 1	14,000	7,000
Hinge rods		
Type 1	2,800	1,400
Type 2	2,800	1,400
Inventory before, 31,400; inventory after, 15,400 (51% reduction).		

showed just how unique each craftsperson was in accomplishing the same task and how the times varied. These times were then documented using load diagram charts. The charts were presented and discussed by the team to develop a standard practice, weed out the time-wasting aspects, and develop job description sheets (JDS) that could be posted at each work station. This enabled workers to reference the procedure during the work day. The JDS established a standard assembly technique that would hopefully improve the quality of the product over time.

Conclusion

It was reported that in addition to the immediate 50% reduction in inventory and 30% reduction in costs, they saw a 5% increase in productivity in the three-quarters of production after the plan was implemented. That improvement has leveled off since January 2011, but management is still very pleased with the results. It appears that the employees are very proud of their clean and organized work stations. The assembly team is especially proud of their contribution to this initiative.

Comment

As mentioned this is an example of a manufacturing plant that has limited if any decision points for choosing alternative materials or processes. Nevertheless they do have significant opportunities to control waste and resources. In this instance waste reduction comes about through a change in manufacturing practice carried out by the employees on the assembly line (the operations function). Support and guidance are provided through the efforts of an industrial engineer (the manufacturing engineering and technology function). The primary technique used is the 5S methodology, a lean manufacturing tool. Once the materials and processes have been established and the product design has gone into production, making use of the techniques of lean manufacturing is the next step in eliminating waste.

CASE STUDY 3

The company in this case study is a Stage 2, Group 2 manufacturer that makes parts to the customer's specifications. They are also specialists; therefore it can be said that there are no decision points available to them. They have limited or no control over the selection of materials and since they are specialists their primary process generally defines the other processes and operations required to complete the item or component. So the question is what approach do they use to minimize waste and resources? Although there is no exact method that must be followed to achieve this goal, these companies do have a lot in common. This case study is an excellent example.

Introduction

Engineered Sintered Components (ESC) is an ISO/TS 16949 supplier to the automobile industry. ISO/TS 16949 is an auto industry technical specification based on ISO 9001. Their plant is also certified under the ISO 14001 environmental standard. This profitable metalworking company has also been recognized as a *Steward of the Environment* by the North Carolina Department of Environment and Natural Resources.

The company is located in the Piedmont region just north of Charlotte, North Carolina. There are two buildings on the site, totaling approximately 174,000 square feet. The larger building (144,000 square feet) contains the presses and ovens used to form automotive parts using powdered metal technology. The smaller of the two buildings is dedicated to secondary machining operations. The company operates on a three-shift basis and employs about 360 people. Annual sales are near $60,000,000 with a plant capacity capable of producing 600 tons of parts per month.

Description of the Manufacturing Operation

When we arrived at the plant the receptionist gave us a "tri-fold pamphlet" outlining the company's safety and environmental requirements for visitors and contractors. She asked that we "study it" while we waited for our hosts in the lobby. It was our first indication of the extent of ESC's commitment to safety and their environmental policy. Once inside we met with managers in charge of Human Resources, Training, and Operations. We had talked with them earlier to explain we were eager to learn how a manufacturing company can be a leader in environmental stewardship and successfully meet the competitive demands of the automobile industry. To answer our questions they started by giving us an overview of the manufacturing operation, followed by a tour of the plant.

Manufacturing Operations and Sequence

Engineered Sintered Components uses a pressing process to form parts by compacting powdered metal in a die set and then heating the part in an oven to just below its melting point to sinter or fuse the powdered metal. The result is a dimensionally accurate part with excellent physical properties. These parts, made from ferrous alloys, are designed to meet some of the most demanding automotive applications. Some examples are engine valve train sprockets and gears, valve guides and seats, power steering components, and antilock-brake sensor rings.

There are not too many steps in creating a part using powdered metal technology. The processing starts with a powder metal stock, a fine powder less coarse than granulated sugar, in pure or preblended form. This stock is purchased from one of three suppliers.

1. *Blending and Weighing.* Powder metal is mixed with a dry lubricant and other additives such as carbon, copper, nickel, and/or other alloying materials. There can be as many as seven items that can be mixed with the base power to create the blend. This blend or batch will weigh 1000 kg (2200 pounds).

2. *Pressing.* An overhead hoist lifts the blended stock and places it in a hopper above a press. The press contains the

Figure 4.2. A green part as it comes from the press.

precision tooling that will compress the powder to create the dimensionally accurate part. Based on part design the press compacts the powdered charge with a force of 3 to 7 tons per square centimeter (19.4 to 45.2 tons per square inch) to create what is termed a "green part." When it is removed from the press it appears to be a finished part. The high pressure creates a mechanical bond between the powders that provides sufficient strength to allow the part to be handled until it is fused. Figure 4.2 shows a part that was just pressed.

3. *Sintering.* Although the green parts are nearly dimensionally correct they have little mechanical strength and can be broken with rough handling. Therefore the parts are moved with care to a controlled-atmosphere furnace. These furnaces heat the part to just below the melting point of the principal material so that the particles are fused, or sintered, creating a component that has the metallurgical properties of the blended material.

Figure 4.3. A completed part before a rust-preventative treatment is applied.

4. *Finishing.* Once the part has been sintered it may require cleaning, surface treatments such as shot blasting, carburizing, or heat treating. In some cases secondary operations such as sizing, coining, repressing, or machining are also required. Figure 4.3 shows a completed part that is ready for packaging.

5. *Preservation Treatments.* Some parts go through a process to create a layer of black oxide (magnetite) on the exposed surfaces of the part. This process exposes the part to high-pressure steam at a temperature just over 900 degrees Fahrenheit. The result is a hard black oxide finish that resists rust.

There can be additional secondary operations such as oil and resin impregnation that may be required to limit porosity or give the part self-lubricating properties or improve machinability. Other work can include machining processes such as drilling, tapping, and brazing.

Steps Taken to Lessen the Environmental Impact of the Manufacturing Facility

After the plant tour we met again to learn how ESC is able to meet their environmental responsibilities while improving their manufacturing operation. Their answer to that question was that being environmentally responsible is not only good manufacturing practice but it also reduces costs. They indicated that identifying waste streams is not a formalized practice in itself. They explained it is part of their *Continuous Improvement* efforts.

In general they use the same classifications or categories for significant waste streams and resources that are on the PCPC. Once they have identified a source for a major waste stream they examine alternatives to reduce or eliminate the source and the associated costs. Some current examples by category are listed in the next subsection.

Material Waste Reduction

Solid Waste

- The one-time-use shipping containers called Super Sacks that hold 1000 kg (2200 lb) of powder were originally intended to be discarded. Now these bags are being recycled and used as a filler material by another company. This contributes significantly to their 60-ton annual reduction of solid waste sent to the landfill.
- Recycle Press Feed Line Powder. A system is in place to collect feed line powder blends after a production lot is completed. The returned powder is identified by blend for use again.

Liquid Waste

- Oils and other working fluids are being changed based on physical or chemical properties instead of "time in service" or calendar schedules. This has reduced the amount of liquid waste and the costs to replace what would otherwise be a serviceable fluid.

• Waste or used lubricants are being collected and stored so they can be recycled to be used as #2 heating oil.

Gaseous Waste

• Lubrication of parts prior to the coining process was done in an open dip tank. Air flow over the tank and the large exposed surface area allowed significant amounts of *volatile organic compounds* (VOCs) to escape into the atmosphere. This dip-tank operation was replaced with a covered tank that limits the lubricants' exposure to the atmosphere.

• *VOC Reduction.* Use of a hydrocarbon cleaner for part washing was eliminated and it was replaced with an aqueous cleaner.

• An open tank system for the application of rust preventatives was replaced with an enclosed system to limit VOC release. Also the original rust preventative was replaced with a product that had a reduced amount of VOCs.

Resource Reduction

Water

• Condensate from refrigeration systems is being collected and used as "mop" water—water for cleaning and scrubbing floors.

• Condensate from chillers and air conditioning systems is being used to replenish water lost in the evaporative cooling system used to cool the plant. Currently 3425 gallons per day is saved.

• A backup well plus condensate water now supplies 50% of all water used.

• An improved water filtration system was put in use to prolong the life of coolant water.

Energy

• The metal halide light fixtures in the plant were replaced with high-efficiency fluorescent lights (T5), which reduced energy consumption from 200,000 kilowatt-hours (kWh) to 82,000 kWh per day.

- Flow meters were installed on the sintering ovens to monitor and regulate natural gas use more closely. This resulted in a 30% reduction in natural gas use.
- An electric preheat stainless steel muffle was replaced with a ceramic muffle, extending the life cycle of the muffle from 1 year to 3–5 years. This resulted in a reduction in natural gas used at the sintering furnace of 5000 therms per month.

Methods the Company Uses to Identify and Make Improvements

The examples listed above are representative of ESC's recent efforts in reducing waste and the use of resources. We next questioned our hosts about the methods ESC uses to identify and implement these improvements. As you might expect the certified management and quality systems (the ISO standards) provided a basis for their approach. They also indicated that their overall strategy is based on the practice of continuous improvement. Beyond that, ESC has some specific programs to support these efforts. The major ones are as follows:

- They have a VOC engineer who works on-site to look for cost savings and value-added opportunities.
- Incentives and recognition are given monthly to associates who make suggestions and submit ideas that improve operations and reduce waste.
- EHS (Environment, Health, and Safety) training is mandatory for all subcontractors before coming on-site.
- Quarterly EMS (Environmental Management System) meetings (for all shifts) have been established.
- Monthly environmental inspections are carried out by employees and management.
- Management provides employees an update each month on the plant's financial performance (profit and loss), operational performance data, and plans for the future.

- Representatives from the company are actively involved in outside organizations and groups to help, support, and demonstrate their commitment to the community. For example:
  ○ Town of Troutman Beautification Day
  ○ Providing internships and education opportunities with Catawba College Environmental Science Education Program
  ○ Establishing a recycling program at Troutman Elementary School
  ○ Participation in Troutman Adopt-a-Highway program
  ○ Participation in local environmental forums
  ○ Membership in the Statesville Chamber of Commerce
  ○ Annual Environmental Stewardship Promotion with local high schools' science classes
  ○ Membership in the Centralina Workforce Development Board

Summary

What is unique about the programs and endeavors at ESC to be a compliant and environmentally responsible manufacturer is their belief that waste reduction and conserving resources is an integral part of what a company must do to be profitable. Since they are a supplier of parts to automobile manufacturers they do not have the opportunity to use several of the tools that are available to companies that have both product design and manufacturing design responsibilities. On occasion ESC does have some limited input on product design but they are primarily restricted to the tools available to the departments of Manufacturing Operations, Training and Development, and Human Resources.

Therefore we found numerous examples of good manufacturing design and practice. During our tour of the plant real-time displays of computer-generated process control, manufacturing lot tracking systems, and other data collection methods were apparent. Our host stopped frequently to ask an ESC associate a specific question about a process or a part. In each case the individual was knowledgeable and detailed in answering the question(s). We also noted that the flow of material through the plant was logical and easy to understand. Also, the plant and equipment were clean and orderly. The one surprising aspect was the large amount of work-in-progress present throughout the plant. Nevertheless there was ample evidence to indicate that this is an exceptional manufacturing operation.

We realize that this description falls short of providing a complete account of the company's operations and all the efforts that make them profitable. Just following the steps we've described would not ensure that a plant making a similar product would be as successful. So a key question might be—what does make the difference? Part of the answer is the people who were involved in our meeting, the persons in charge of human resources, training and development, and manufacturing operations. They are the representatives of the three significant functions responsible for operating a profitable and environmentally compliant manufacturing plant.

Comment

This case is an example of a Stage 2, Group 2 manufacturer, a specialist. Nevertheless this company's approach shares some similarities with the two previous case studies. In each case the companies have subscribed to a certified management system and in the examples given most of the steps they have taken to be environmentally compliant have also resulted in a reduction in operational costs.

CASE STUDY 4

This case study looks at another Stage 2 manufacturer, Corning Cable Systems, LLC, located in Hickory, North Carolina. The facility, officially known as the Hickory Manufacturing and Technology Center (HMTC), designs and manufactures indoor and outdoor fiber-optic cable. The glass fiber used in the cable is made from very pure silica glass in diameters not much greater than a human hair. These fibers serve as waveguides or "light pipes" that transmit light signals and can be bundled to carry video images. In data transmission service fiber optics have less loss over longer distances, are immune to electromagnetic interference and cross talk, and can operate at higher data rates than competing methods of transmission. Consequently, optical fibers have largely replaced metal wires in many digital communication transmission applications.

Introduction

Corning's Hickory, North Carolina, plant opened in 1982 and has manufactured and shipped more than 58 million miles of fiber-optic cable since it began operations. The plant produces a variety of cables, some of which are shown in Figure 4.4. The applications for these cables fall into two primary markets, enterprise networks (i.e., data centers, government, education, and

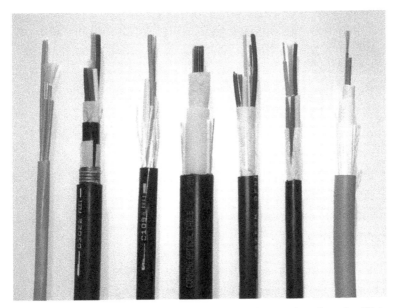

Figure 4.4. A few examples of the fiber-optic cables produced at the Hickory plant.

Fortune 500 companies) and carrier networks (i.e., telephone and cable television companies).

We met with two representatives from HMTC; one is responsible for environmental programs and the other has responsibilities in the department that manufactures connectors for fiber-optic cable. We began by noting the site's recognition by the state of North Carolina as an "Environmental Steward." North Carolina's Department of Environment and Natural Resources (NCDENR) stated that Corning's Hickory facility "has demonstrated environmental leadership through its commitment to exemplary environmental performance beyond what is required by regulation." The commentary also contained the NCDENR's advisory board's reasons for selecting Corning's Hickory site as an Environmental Steward. They indicated the company's:

- Integration of environmental management into core business operations through multiple processes for internal communication of environmental issues and driving environmental awareness through efficiency improvement and awards programs
- Practice of operating well below permitted limits and implementing management techniques beyond those required by regulation
- Reduction from 2006 to 2008 in its energy usage by 22 percent, normalized to production
- Increase in its recycling of plastic material in 2007 by 17 percent from the 2006 level
- Decrease in water consumption. During the three-year period beginning in 2004 the facility decreased its water consumption by approximately 59 percent when normalized to production.

The Corning representatives commented that all of these rates are normalized to a per unit basis. The point being that fluctuating production levels require performance metrics to be based on a unit of production. Specifically, energy use and water consumption are two environmental aspects measured by this method. It was also pointed out to us that achievements cited by NCDENR were the result of efforts that began with the site's ISO 14001 certification several years earlier.

The Hickory Manufacturing and Technology Center obtained ISO 14001 certification in November of 2000, at which time the environmental management system (EMS) was implemented. A fundamental characteristic of HMTC's EMS is the requirement for continuous improvement of its manufacturing operations and environmental performance. This has been key to reducing the site's environmental impact and improving the long-term financial success of the operation. Every employee, as well as the plant's management, is actively engaged in continuous improvement. This was emphasized in our conversation and it was pointed out that continuous improvement is the "force" that earned the plant the title of a Steward of the Environment.

Manufacturing Operations and Sequence

As previously mentioned, the primary products manufactured at this facility are data transmission cables containing glass fibers. As you can see in Figure 4.4 these cables are fairly complex and may contain several bundles of optical fibers. The actual fiber comes from other Corning plants in North Carolina. The Hickory plant essentially "identifies" the fibers and then "packages" them in a manner that provides protection from mechanical and environmental damage.

The processes used to produce an outdoor fiber-optic cable will give you an idea of the manufacturing sequence. Figure 4.5 illustrates the cable components that are referred to in the following steps:

- The first process is color coding the individual glass fibers using a UV-curable ink.
- The color-coded fiber then moves to the ribboning process.
- The ribboning process coats the fibers to form a thin flat "ribbon" profile. A ribbon can contain between 12 and 36

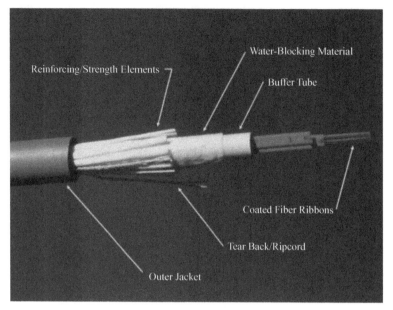

Figure 4.5. The components found in an indoor fiber-optic cable.

optical fibers. Each ribbon is printed with a number and color code for identification purposes.

- The ribbon then moves to the next process, where it is placed inside an extruded plastic tube. This is called a buffer tube and it protects the fiber and/or ribbons. A buffer tube may contain between 12 and 864 fibers.
- The tube is then processed into a cable where strength elements are added along with water-blocking materials. The reinforcing or strength elements include flexible fibrous polymers along with one or more ripcords to aid in removing a portion of the outer jacket. Finally an outer jacket sheath is applied to cover the cable.
- The final cable is spooled onto a reel and shipped to the installation site.

The jacket and buffering tubes are created using a continuous extrusion process. Continuous extrusion enables Corning to produce cables up to 12 kilometers long. Polyethylene, typically black, is used to provide long-term UV protection to cables exposed to sunlight. In some outdoor applications the cables are wrapped in corrugated steel before they are covered with a protective plastic sheath. The armor aids in protecting the glass fiber from damage during installation and from rodents once installed.

Steps Taken to Lessen the Environmental Impact of the Facility

Corning Incorporated, the parent of Corning Cable Systems, LLC, developed an EMS to fit the needs of most of its facilities. The approach that was used can be described as a pyramid. Figure 4.6 is a depiction of that pyramid.

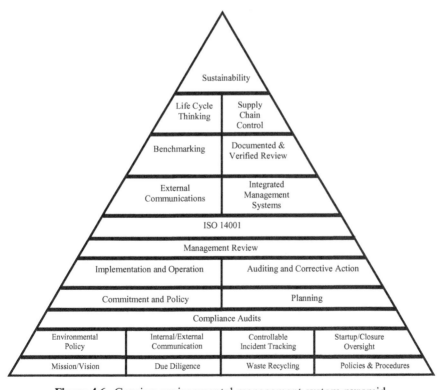

Figure 4.6. Corning environmental management system pyramid.

The components of the pyramid identify essential elements of the management system; however, every one of Corning's companies has to integrate these elements into its own management approach. The Hickory plant has accomplished this by incorporating some basic lean manufacturing techniques.

Methods Used to Make Improvements

The lean manufacturing methods and techniques used by HMTC to reduce waste are "actionable" techniques. The focus is aimed at methods and techniques that manufacturing associates (employees) can use. Kaizen was a term used frequently in our discussion. It is a foundational method in HMTC's approach to lean manufacturing.

Kaizen is the Japanese word for "improvement" or "change for the better". The word has come to describe the practices that focus on continuous improvement of manufacturing processes and operations. Kaizen as practiced at HMTC includes the following features:

- All associates should realize that continuous improvement is a way of doing work. In practice they should continually seek ways to improve their own performance.
- Make small changes rather than large far-reaching changes. Continually making small advances can lead to significant improvements over time.
- Small improvements or incremental change are usually less expensive.

The Kaizen program at HMTC is an incentive-based program that is more than a suggestion system. Every associate on the manufacturing floor is expected to be involved in continuous improvements. In practice this means each person is providing ways to improve the manufacture of the product. Furthermore, each individual that submits an improvement must also be involved in the implementation process. The Kaizen program also extends to environmental improvements.

In our conversation several other lean manufacturing methods were mentioned. The following are some examples of tools that are used as a means to achieve Kaizen. The ones listed here also help to illustrate the range of methods and the variety of the approaches used to achieve continuous improvement.

The "*5S*" process refers to the Japanese words Seiri, Seiton, Seiso, Seiketsu, and Shitsuke. A translation is Sort, Set in Order, Shine, Standardize, and Sustain. These symbolize the five simple steps to create a better working environment. This method has been well used by the manufacturing associates to make improvements.

The *Five Fundamental Elements of Lean Manufacturing* also serves as a guide for establishing a lean production facility. In practice these elements can be considered goals. The associates work to achieve:

- Uninterrupted flow of product
- Fast changeovers (single-minute exchange of dies)
- Error-free processing
- Better workplace organization
- An ongoing drive to improve

There is another approach in use called *DMAIC*, which stands for "define, measure, analyze, improve, and control." It is a methodology for solving problems and more. DMAIC is used as a means to acquire a better understanding of the process and how it should be operated and controlled. The steps are as follows:

1. Define the problem.
2. Measure key aspects of the current process and collect relevant data.
3. Analyze the data to investigate and determine the root cause of the problem.
4. Improve or optimize the current process using techniques such as design of experiments, *poka-yoke* (which means

mistake-proofing), and decision-point analysis. Confirm the improvement.

5. Control the process or operation and determine if all the significant independent variables are known. Confirm using statistical analysis.

There are other lean manufacturing techniques in use at the plant but the ones listed here are representative and serve to sketch out their approach. It is important to note that these techniques are intended to involve all employees from management to the manufacturing associates. The essence of the system is continuous measurable improvement. As objective evidence the plant maintains the metrics to show the magnitude of the improvements that have been made in waste and resource reduction.

Examples of Waste Reduction

Employees at HMTC take environmental stewardship seriously. For several years the plant had a practice of recycling standard items such as cardboard and scrap metal, but the plant expanded its recycling program and encouraged individual employees to submit ways to either eliminate or recycle other forms of waste. In 2008, the plant recycled more than 45,000 pounds of paper—four times the amount recycled in 2004. Since the program began, thousands of improvements have been implemented by employees who now collect, segregate, and ship out more than 80 different types of materials, diverting more than 6.5 million pounds of waste from the local landfill.

There have been reductions in emissions as well. Direct greenhouse gas emissions have been reduced 83% from the peak year of 2008.

Energy consumption at the facility has also been reduced. One example given demonstrated that significant savings can be found that does not directly involve the processing equipment in the plant. In this instance the ventilation system in the compressor room was changed so the air being exhausted was not replaced with air-conditioned or heated air from within the building. The expected energy savings during the winter is 589,000 kWh and the summer savings are estimated at 736,500 kWh. Another project involved the installation of three mass airflow handlers for manufacturing areas. The projected energy savings is 574,000 kWh.

Comment

Corning Incorporated has established a corporate goal of sustainability based on ISO 14001 principles. The EMS is intended to be integral to the operational management of its individual manufacturing plants. For a manufacturing plant, sustainability means being environmentally responsible and profitable. As a means to accomplish this goal the HMTC has established the practice of lean manufacturing, and Kaizen is at the center of the approach.

There is evidence the plant is reaching its goal. The examples cited in this case study are just examples and are not intended to be prescriptive in the sense of indicating what to do. The essence and the truly innovative aspect in this case study is the company's method for getting everyone involved in continuous improvement. There is an excitement factor in the approach that is refreshingly unique.

CASE STUDY 5

Introduction

This case is based on a conversation one of the authors had with Hong Yang, Manager of the Waste Minimization and Recycling Group, which is part of the Waste and Resource Management Department in Singapore's National Environment Agency (NEA). It was her group that published the Waste Minimization Guide cited earlier in this chapter. Much of her group's work deals with the waste that is generated in the third and final stage of manufacturing. Her comments and thoughts provide some insights to the opportunities and challenges that companies and communities face in their efforts to reduce waste and minimize resources at this stage of the manufacturing sequence.

As you recall, this stage of manufacturing involves the distribution of products, their sale, service, and disposal. Except for vertically integrated companies most manufacturers are not directly involved in this final stage of manufacturing but there are a few exceptions. One exception is the packaging that companies provide to ship their products or components to distributors. Another involves producers of consumer products that provide point-of-sale packaging. Although most manufacturers have a limited role in reducing waste in this stage there are some significant opportunities.

For Singapore the generation and disposal of solid waste is a critical problem. Singapore is one of the most densely populated countries in the world. It is a city-state of 712 square kilometers, or about one-fourth the size of the state of Rhode Island, and it has a population of about 5.2 million people (see http://www.singstat.gov.sg/stats/keyind.html). In 2010 the International Monetary Fund ranked Singapore 39th out of 183 countries in terms of the value of its gross domestic product.

Back in 1970 their vigorous economy was generating 1200 tons of solid waste that was landfilled each day. The dwindling landfill space due to the nation's land constraints meant that other methods had to be found to supplement the landfills. In the early 1970s Singapore made a decision to construct waste-to-energy incinerators. Incinerating refuse offered a volume reduction of up to 90%. Despite this, by 1999, Singapore was no longer able to landfill waste on its main island and had to resort to building an offshore landfill. By 2000 the amount of refuse to be handled had increased to 7600 tons per day in tandem with economic and population growths.

Efforts were intensified to reduce waste through the 3Rs (Reduce, Reuse, Recycle) program. This helped to put off building a new 3000-ton-per-day waste-to-energy incinerator once every 5 to 7 years to once every 10 to 15 years.

In our discussion Ms. Yang described several of the initiatives that have been implemented. The waste-minimization guidebook is one example. However, there are other programs that have been put in place. Together they are producing results. In 2006 the amount of solid waste disposed of fell to 7000 tons per day. This was despite a growing population (increasing from 4 million in 2000 to 4.4 million in 2006) as well as higher GDP. This is a 600-ton-per-day savings compared with the waste disposed in 2000. A more recent initiative to reduce waste at source is the "Singapore Packaging Agreement."

Singapore Packaging Agreement

Prior to the launch of this agreement the NEA reviewed the packaging polices of several countries before they drafted their program. The agreement, which is voluntary, is modeled after New Zealand's Packaging Accord along with some features of Australia's National Packaging Covenant. The implementation of the Singapore Packaging Agreement is overseen by a governing board that comprises representatives from government, industry, and interested nongovernmental organizations (NGOs). The NGOs have the task of helping to educate the community and citizens on the importance and ways of minimizing packaging waste. A key aspect of the agreement is its nonprescriptive approach that offers more flexibility to manufacturers, importers, and retailers to develop workable cost-effective solutions for reducing packaging waste. At the launch in 2007, 32 organizations including industry associations, companies, public waste collectors, and NGOs had signed the agreement. As of January 2012, 128 organizations are participating in this program. The companies involved range from a fast food chain to a computer brand owner.

Ms. Yang pointed out that corporate and industry support is vital for a voluntary program to succeed. An important aspect of the program is the recognition a participating company can receive in the form of an award. The awards certainly honor the recipient but they also serve to identify best practices and motivate companies to participate and to continue their efforts to improve. However, an equally important aspect of the program is to get the end user, the consumer, to participate in waste reduction.

The "Three Rs" Approach to Solid Waste Minimization

Most of the packaging that is used to facilitate distribution while protecting the products from damage is single use. A major method being promoted to reduce packaging waste embodies the concept of the "Three Rs" for minimizing waste:

- Recycle
- Reuse
- Reduce

In Singapore this concept of recycling is well known and many companies have established recycling programs. However, this is only a first step. Recycling single-use packaging has a significant drawback. The recycled material reverts to being a raw material for creating a "stock." This means it has to start again at the first stage of the manufacturing sequence and then progress through the second and third stages to once again be distributed for use as a product.

Reuse is a more effective means to reduce waste and the use of resources. An example that is often used is comparing a single-use plastic drinking cup with a glass cup. The glass cup requires washing after use while the plastic cup, if recycled, must be transported back to Stage 1. The reused item avoids the resources required to convert the recycled item into a useable manufacturing stock.

The third R, reduce, is the step that provides a significant opportunity to minimize waste and resources. Consider this step as a challenge to those that design the product, manufacture it, sell it, use it, service it, and then ultimately dispose of it. Ms. Yang used household appliances as an example of the challenge and the opportunity that exist for manufacturers. "Today the concept of reparability is not as prevalent as it once was." She noted that improving reliability of products is a significant component in minimizing waste.

However, repairability of products is often ignored. She used a handheld hair dryer as an example. This small appliance uses resistance elements to heat the air being propelled by a fan motor. The serviceability of the hair dryer is largely dependent on the life of the heating elements, switch, and the fan motor. To keep the initial selling price reasonable, the service life of these three items under normal use is much less than the dryer body and the cord set. Consequently when any one of the three electrical components fails the hair dryer is discarded. The point is that manufacturers typically don't view small appliances to be repairable. But, if they were repairable, it would require an infrastructure to provide parts and service at an economical cost; none of this exists in many parts of the world. This also implies that consumers would be willing to take the time and effort to repair or seek service to repair their household appliances. Repairability (servicing) rather than discarding can reduce waste.

There are economic implications to manufacturers that shift from "discard/replace" to "repair/reuse." In part this loss of sales volume generated by the discard/replace approach can be offset by offering economical repair parts and/or a network of repair/ rebuild centers. This would be feasible if first the product is technically current and as mentioned consumer attitudes favor extending the life of their current products as opposed to discarding and purchasing a replacement.

But she also made the point that there are circumstances where a product should be discarded rather than serviced to extend its life. One example would be an appliance that incorporates improvements in technology that makes it significantly more efficient or lessens its environmental impact. The automobile industry is another example. Automobile technology, particularly in the past decade, has provided some major improvements in safety, reducing emissions, and decreasing fuel consumption. It is questionable whether an auto built 10 years ago should be maintained in service. However, notwithstanding these exceptions, the concept of repairing to extend service life is compatible in spirit with "Reuse" and "Reduce."

Comment

Stage 3 manufacturing provides a more diverse situation for minimizing waste and resources since the number of organizations involved in distribution, sales, service, and disposal are numerous and seldom if ever connected under a single management system. It can be reasoned that if there is one management system it is the government. Consequently this case provides that perspective. A key aspect, however, is the government's management role in representing and bringing together all the groups involved in waste minimization. In Singapore it is called the People, Private, and Public initiative, the 3P initiative. This initiative encourages these three sectors to work together to develop innovative and sustainable environmental solutions for minimizing waste.

CONCLUSION

In these case studies the approaches used to reduce waste in manufacturing vary to some extent depending on the industry and the type of manufacturing facility. However, there are some commonalities. First, certified management systems are prevalent in companies that are doing well in being environmentally responsible and profitable. Second, it is not an unusual approach for manufacturing companies to subscribe to their industry's "best practices." However, coupled with best practice are the concepts of lean manufacturing. This is a philosophy of operation that we found to be well known and widely practiced. Finally, these companies frequently participate in voluntary governmental or industry-sponsored environmental initiative programs.

SELECTED BIBLIOGRAPHY

American Chemistry Council (ACC). Responsible Care guiding principles. Available at http://responsiblecare.americanchemistry.com/Responsible-Care-Program-Elements/Guiding-Principles/default.aspx (accessed February 9, 2012).

International Council of Chemical Associations (ICCA). Responsible Care®. Available at http://www.icca-chem.org/en/Home/Responsible-care/ (accessed February 9, 2012).

NEA (2003). *Guidebook on Waste Minimization for Industries*. Singapore: National Environment Agency.

NEA (2007). Singapore Packaging Agreement. Singapore: National Environment Agency. Available at http://app2.nea.gov.sg/topics_packagreement.aspx (accessed April 20, 2012).

NEA (2010). Chevron Oronite Pte Ltd. Singapore: National Environment Agency Available at http://app2.nea.gov.sg/data/cmsresource/20100901491845260649.pdf (accessed February 9, 2012).

NEA (2011). 3R Packaging Awards 2011. Singapore: National Environment Agency. Available at http://cms.nea.gov.sg/data/cmsresource/20111027672361400348.pdf (accessed April 20, 2012).

CHAPTER 5

AN OVERVIEW OF TOOLS USED TO IMPROVE MANUFACTURING OPERATIONS

INTRODUCTION

In the first chapter there were several techniques and concepts mentioned that manufacturers can and are using to be profitable and minimize waste. It helps to think about these methods as basic techniques or tools since they are not industry specific. In this chapter we examine a few more of these tools and classify them according to the functions or groups of people within the organization that could make use of them.

Recall that in Chapter 1 there were three key functions—marketing, product engineering, and manufacturing engineering—that define the product and how it will be manufactured (Fig. 5.1). Once a product's characteristics have been established, the materials specified, and the processes selected these three functions have had an opportunity to significantly reduce the potential for creating waste.

Improving Profitability Through Green Manufacturing: Creating a Profitable and Environmentally Compliant Manufacturing Facility, First Edition.
David R. Hillis and J. Barry DuVall.
© 2012 John Wiley & Sons, Inc. Published 2012 by John Wiley & Sons, Inc.

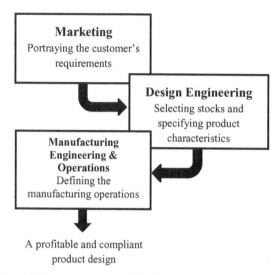

Figure 5.1. The initial group responsible for creating a profitable and environmentally compliant product.

However, as soon as the first lot of products is manufactured a new group will take over the responsibility for minimizing waste. They too have plenty of opportunities to reduce waste. This second group consists of manufacturing operations, human resources, and training (education) and development. Figure 5.2 portrays the relationship these functions have with each other.

Contained within each of these functions are a number of traditional professions, work groups, or disciplines. For example, "manufacturing operations" includes manufacturing engineering, industrial engineering, plant engineering, factory floor managers and supervisors, and production employees. The titles of these three functions therefore are just general markers or buckets to recognize who is responsible for waste reduction and who would be more likely to employ these tools.

By now you may have realized that waste reduction and minimization is a two-step approach. In the first step there is a major opportunity to reduce waste when designing a new product and being able to select processes to produce the new part. Once the design is established and/or the processes are in place the

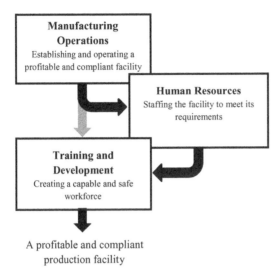

Figure 5.2. The second group responsible for operating a profitable and environmentally compliant manufacturing facility.

responsibility shifts, but not entirely, from the marketing and product engineering group to the second group. There are always revisions to every design, which briefly puts the responsibility back onto the design group. Regardless of which group is involved, first or second, each has a wide range of tools that can be used to control and/or reduce waste.

Before we look at these tools let's examine a technique that can be used by both groups to establish a basis for waste reduction that is both environmentally sound and profitable. This technique is often used by accountants and bankers. Wait, it does have some relevance. In the financial world it's called *sources and uses*. When an accounting firm first takes on a company as a client they want to understand where the company's revenue comes from and what it is used for: the sources and uses. This helps them understand the dynamics of the business.

So an organization that wants to reduce waste must first understand the source of all the materials and the energy and resources (recall the major headings on the profitable and complaint process chart) and how they are used in the facility. Identifying the major

Figure 5.3. National Graduate Institute for Policy Studies in Tokyo, Japan.

materials and resources and what they are used for becomes the beginning "inventory" for the waste reduction effort. It also helps to raise questions about alternative sources that would reduce waste or lessen the environmental impact of the operation. An example here will help to explain this approach.

The National Graduate Institute for Policy Studies in Tokyo, Japan, is a graduate school whose goal is to advance interdisciplinary policy research that relates to the real world. The result of the institute's research is a scientific approach used to train administrative officials and policy analysts. The institute is housed in a modern multistory building shown in Figure 5.3.

In the lobby of the institute is a flat panel video display showing information about conference room use and data on electric power consumption. The data listed is kilowatts per square meter, kilowatts, kilowatt-hours, and kilograms of CO_2 produced. Along with the display is a diagram that indicates where the facility can obtain power. A solar panel array, which can be seen on the roof of the building in Figure 5.3, is one of the sources. Figure 5.4 shows the display and the diagram.

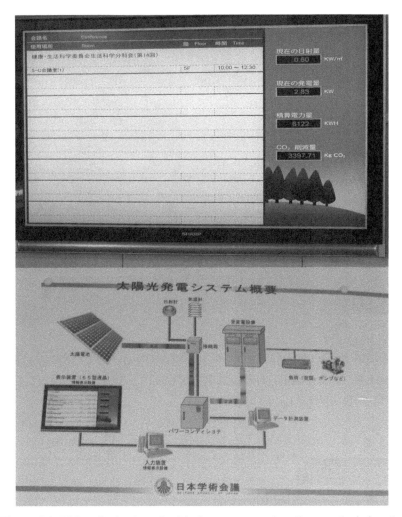

Figure 5.4. Video display board, with the accompanying diagram depicting the sources of electrical power being used in the building.

Placing the display panel in the lobby makes the information available to casual visitors as well as to those involved with the institute. This is a terrific way to create awareness.

Getting back to the sources-and-uses concept: creating a listing of the materials and resources being used in a facility provides a starting point. The next step of course is to relate the materials and resources used to an activity.

As an aside, there are a number of technologies that have been called "green." Often there is pressure to adopt these technologies without evaluating their potential to reduce waste. The capital investment needed to install and operate these technologies has to be evaluated in terms of the amount of waste that can be reduced compared with other investments. The objective is to reduce the amount of waste generated, whether it is CO_2 or VOC emissions, not showcasing popular "green" technologies.

A complete sources-and-uses listing is usually a matter of pulling information together. In some cases a company may have to invest in smart metering to identify where electrical power or water is being used. But most companies can identify the significant users by using the equipment "nameplate" specifications and operation times. Therefore the next tool to be used is the profitable and compliant process chart.

WASTE REDUCTION: THE PROFITABLE AND COMPLIANT PROCESS CHART—A COLLABORATIVE TOOL FOR BOTH GROUPS AND ALL FUNCTIONS

This is the fundamental tool that can be used by all functions in an organization. We have developed the profitable and compliant process chart (PCPC) introduced in Chapter 2 as a methodology and decision-making tool. For new products, marketing, design engineering, and manufacturing engineering can use this methodology as a basis for collaboration to create a product design and manufacturing method that reduces or minimizes waste and provides a profitable product for their customers.

Once the product is in production the PCPC can be used by the second group to continue working toward the objective of waste reduction. For this group many if not all of the product characteristics are fixed so the product design and marketing groups have a minimal role to play. However, finding opportunities to reduce waste is an ongoing responsibility or, to put it in more positive terms, an ongoing opportunity. As you may recall, the case study of Engineered Sintered Components (ESC; Chapter 4, Case Study 3) provided several examples of significant waste reduction that was achieved with established product designs.

The tools listed in the next section are organized by groups; however, that does not imply that they are restricted to that particular group. In many cases they have universal utility. For those of you who have used a particular tool, you'll recognize that our description falls short of providing a comprehensive explanation of its use. That's because this chapter is a guide; you'll need to find other references that are more comprehensive.

TRAINING AND DEVELOPMENT

Training and development is an important function in the second group, the people who produce the products. The individuals in this functional area of the business are responsible for providing the specific information and skills an organization needs to carry out the activities of manufacturing. At one time this function was based on "on-the-job" training. Today, however, most companies have formal training and development departments. One of the most universal skills this department can teach is the concept of operator self-control.

Operator Self-Control

This is the basic building block for management. Joseph M. Juran and Frank Gryna developed this concept when the United States was struggling to overcome the problems plaguing their attempts to place satellites in orbit. *Operator self-control* defines the fundamental work objectives for every individual in the business of manufacturing. The concept's precepts can be stated very simply. After receiving training for their job an operator will be able to recognize or know:

- When they have the correct materials to do their work
- When they have the correct tools and/or their machine is set up and operating properly
- How to do their work correctly
- When their work is not correct and how to to take corrective action
- How to get help if they are not able to correct the problem

Culture Change

One of the most difficult challenges for training and development professionals is cultural change. Making waste reduction a part of the culture of manufacturing is another basic building block in the creation of a profitable and environmentally compliant manufacturing operation. It can be done; reducing waste as a way of life is taking hold in many countries. For example, we are all aware of the need to recycle. Separating household waste into recyclables and landfill waste is a widespread practice in the United States. However, some countries have recycling as a fundamental part of their culture. In Japan waste receptacles such as the one shown in Figure 5.5 are in everyday use in businesses, fast food restaurants, shopping malls, as so forth.

Actually being conscientious about recycling is a just a first step. The objective is eliminating the need to recycle. OK, that's not really an attainable objective but a manufacturing operation can do much to eliminate the amount of material that needs to be recycled. As an example, more and more companies are reducing the amount of "single-use" packaging material. In Chapter 4, Case Study 5 introduced the "Three Rs" concept of Recycle, Reuse, and Reduce. This is particularly appropriate for Stage 3 manufacturing. For Stage 2 manufacturing the gold standard for waste reduction is quality improvement. Eliminating rejects eliminates waste and the need to rework, another form of recycling.

The point is that eliminating waste is a way of life inside and outside the workplace. It has been pointed out that "good work habits" don't just happen. When a company is considering building

Figure 5.5. A typical waste receptacle unit in Japan. It illustrates how recycling is an everyday practice.

a new facility one of the first questions to be asked is "is there a suitable workforce available?" If the answer is yes, then the next question to be asked is "what are the workforce needs for education and training?" A successful manufacturer must have a capable workforce that has the work habits, knowledge, and skills to perform the tasks required to profitably make a product that is fit for the use of the customer. Today a manufacturer must also develop a workforce that has a concern and a willingness to reduce waste. This is similar to the responsibility that a company has to create a workforce that is safe and safety conscious. The programs and tools used to create and maintain a safe workplace can also be used to instill the awareness and culture to strive for the elimination of waste.

MANUFACTURING ENGINEERING AND OPERATIONS

Manufacturing engineering as stated earlier includes a variety of professional titles and practitioners. Examples are industrial engineers, manufacturing technologists, production engineers, and quality engineers. Operations include supervisory staff, plant management, and the people who actually operate the equipment, and processes to manufacture the product. This section describes methods and tools that all of these people could use to reduce waste.

Lean Manufacturing

This term embodies a range of activities. In the 1980s the term *world class manufacturing* was used to define manufacturing systems that pared away any activity that did not add value to the product being produced. This practice evolved to the elimination of wasted effort, time, and material that did not create value for the customer. Product that was not being worked on or on its way to the next operation that would add value to it was considered wasteful. Idle product is *work-in-process* (WIP) that is tying up resources. The concept of *continuous flow manufacturing* is aimed at minimizing WIP and reducing lead times, thus making the manufacturing system more reactive to the customer. *Just-in-time* (JIT) manufacturing meant that a company didn't maintain a large inventory of materials. JIT was intended to foster the creation of more responsive operations across all stages of manufacturing. Reducing inventories and WIP meant that when quality problems occurred or suppliers missed delivery schedules there was no alternate work available, which meant production stopped. The work stoppage might have been covered up before because of high levels of WIP or "emergency inventories" or "bench stocks." But the JIT concept meant a company could no longer ignore these interruptions; they had to eliminate the problems. The result was that lean manufacturing required manufacturers and their suppliers to dramatically improve quality and the reliability of their delivery performance. Problem-solving skills also had to be developed at all levels of the manufacturing operation. In total all of these efforts eliminated waste and made the manufacturing operation more customer responsive.

The best description I've heard of the JIT concept came from an injection molding department supervisor. He observed "that we don't have any extra material or time to play around with. It really puts the pain back in the manufacturing system, it just hurts too much when something goes wrong and everyone knows right away when something goes wrong."

The need to eliminate defects and the problems they cause brought about the *six sigma* approach to manufacturing. It is a way of thinking using a system of skills and methods to create a manufacturing operation that is very close to being able to produce defect-free products. The term itself, six sigma, refers to a manufacturing process that is in control and subject only to normal variability. The degree of *process control* is such that six standard deviations separate the process mean from each specification limit. Figure 5.6 shows a graphical illustration of a process that would meet this criterion.

This six-sigma process would produce 99.99966% of all of its parts within the specification limits. If you spent the day measuring a specific attribute created by the process and putting the data into

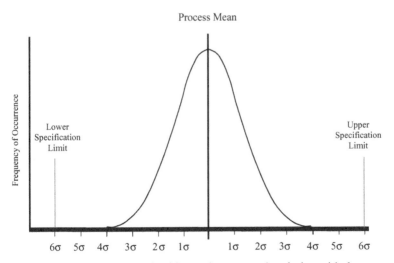

Figure 5.6. A process in control, subject only to natural variation with the process mean six standard deviations from the specification limits.

a histogram, you would find that the shape of the data approximates a normal curve. The distribution (the shape) of the histogram would be symmetrical and tend to hug the mean. Calculating the mean and standard deviation is easily done by entering the data into an engineering or business calculator. Once you have this information place a mark on the x-axis of the plot you created to represent the mean. Then measure six standard deviations from either side of the mean and mark these points. Finally you would need to determine what the specification limits are for the attribute you are measuring. Mark these limits on the plot. If the process is six-sigma capable, the specification limits will fall on or outside the points marking six standard deviations from the mean.

An example will help illustrate this idea. Look back again at Figure 5.6. Notice that the curve representing the number of parts produced for a specific dimension at four standard deviations from the mean is nearly zero. Imagine a machine producing a part that has an engineering specification stating that the diameter must be 1.000 inch plus or minus 0.006 inches. If this is a six-sigma process, then the process mean will be one inch and the standard deviation of the population of parts that the machine produces should be 0.001 inch. Those of you who are familiar with the normal distribution will realize that about two out of every three parts will fall between 0.999 inches and 1.001 inches. This is a very tightly controlled process, which is at the heart of the six-sigma concept.

Incidentally the term six sigma is credited to Bill Smith, an engineer at Motorola. If you do a search for six sigma you'll find companies and universities offering six sigma training and certification at various levels. The concept has developed into a formalized approach to process control and improvement. However, for a manufacturer to achieve this level of control there are other skills and methods needed.

To identify these tools it helps to follow the progression of their development. The concept of natural (normal) variability was first recognized by Walter Shewhart, who was working at Western Electric's Hawthorne Works near Chicago, Illinois, in the 1920s. Statistical process control evolved from his work.

After World War II, W. Edwards Deming went to Japan to teach statistical process control and help the Japanese rebuild their manufacturing economy. He worked with Japanese companies to apply statistical methods to production along with his vision of what is now recognized as quality-based management. Later on in his work he developed his Fourteen Points; these principles go beyond the concepts of process control and define the basis for *total quality management.*

Joseph M. Juran also worked in Japan after the war but independently of Deming. He worked with the Japanese to change the organizational culture that impedes the introduction of change. Juran furthered the concept of quality management as an individual responsibility with the introduction of operator self-control (discussed earlier in this chapter).

Establishing six sigma process control and the other techniques cited are just a part of the concept of lean manufacturing. The concepts of continuous flow and JIT are also useful in reducing waste but, more important, they stress the manufacturing operation to expose problems. What is next are some of the more universal methods used for solving problems and improving operations.

Kaizen

Kaizen is a Japanese word meaning "improvement" or "change for the better." It has become more widely known as the process of *continuous improvement.* In practice it has been used as the basis for reducing costs and improving service in virtually every form of business activity. Continuous improvement efforts usually involve all functions and employees in an organization. It will involve the use of virtually all the tools listed here. The Corning Hickory plant (Chapter 4, Case Study 4) has Kaizen as a foundational approach for waste reduction and manufacturing improvements. The case study provides a useful explanation of this technique and its application.

Pareto Principle

Another contribution of Juran's quality management approach was his use of the *Pareto principle* as a means for prioritizing problems. This was discussed in Chapter 1 but it is worth revisiting Pareto's observations once again. Vilfredo Pareto was an Italian economist. In the early part of the twentieth century he observed that there always seemed to be a disproportionate allocation in the ownership of land. He noted that 80 percent of the land was owned by only 20 percent of the population, the 80/20 rule. Juran found that this disproportionate allocation seems to occur frequently in manufacturing. Some examples are rework costs, defects, and power consumption. In application the Pareto principle indicates that one should identify the "significant few." As an illustration suppose that you are initiating a program to reduce the cost and use of electricity in your plant. It is likely that most of the cost can assigned to a few key processes or uses. These processes would be the significant few that should be identified and examined first. The rest would be the trivial many that account for the remainder of the monthly cost.

Process Control

Statistics and data gathering are often considered to be the basis for statistical process control. However, the terms consistency, repeatability, and predictability are the key aspects of control. As a way to make this point consider this example of process control. You may have driven by or possibly seen photos of corn fields in Illinois, Iowa, and Indiana. If you have, they may have looked similar to the photo in Figure 5.7.

The uniformity in height (disregarding the topography or contour of the field) is the result of the farmer's skill. The tree line in the background provides an example of an "uncontrolled process." The point being that achieving process control provides uniformity in performance. Predictability—being able to predict the exact height of the corn at a specific date—is probably outside the farmer's ability to control because of weather. However, a

Figure 5.7. A cornfield in Illinois.

manufacturer does, in most cases, have the ability to control the variables that affect a manufacturing process. Therefore consistency, repeatability, and predictability are possible.

Certified Management Systems

The certification of management systems formalizes the way a business operates. The ISO 9001 certification in particular implies that the organization has a documented quality management system. A manufacturer that has this certification can demonstrate to its customers that it is able (but it is not a guarantee) to provide quality products. It also establishes a basis for training employee-associates by having documented procedures for completing tasks. It can be a precursor for establishing operator self-control.

Design of Experiments

The *design of experiments* (DOE) technique is used to determine the significant variables that affect a specific outcome in a process or operation. Once these significant effects are identified it enables the operators to control and reduce the variability of the process. This approach to experimentation allows for the investigation of several variables at once using the DOE technique. It also highlights the presence of interaction between variables that is often one of the most confounding problems in manufacturing.

A technique associated with DOE is *DMAIC*, which stands for "define, measure, analyze, improve, and control," the names of the five steps that one would follow in its application. It is one of several methods used to design and carry out an experiment. The objective of a DOE investigation is to acquire a better understanding of the process and how it should be operated and controlled. Using DMAIC the steps to be followed are:

1. Define the problem.
2. Measure key aspects of the current process and collect relevant data.
3. Analyze the data to investigate and determine the root cause of the problem.
4. Improve or optimize the current process. Confirm the improvement.
5. Control the process or operation and determine if all the independent variables are known and confirm by using residual analysis.

Certainly this problem-solving process is not limited to the use of DOE as a technique. Some problems don't require the use of experimentation to find a solution. Simpler techniques in many cases are better. Nevertheless DOE is a powerful tool and the brevity of this explanation shouldn't be construed as minimizing its importance or usefulness.

Poka-Yoke

Poka-yoke is a Japanese term meaning "fail-safe" or "mistake-proofing." Application of this concept to processes, machines, and work sequences means they would be designed to eliminate the possibility of making a mistake. A good example of mistake-proofing is the desktop personal computer. When you remove the case to install an expansion card or a memory module you'll find it can only be inserted into a computer motherboard the correct way.

Finding the Root Cause of a Problem

The importance of lean manufacturing and six-sigma methodologies has made effective problem solving an important skill. A useful way to determine the cause of a problem is the *five whys* method. Shigeo Shingo, an industrial engineer working for Toyota (he wrote about the Toyota production system) popularized this technique. An example shows how it works.

Defective parts are being produced by the assembly machine. A person goes out in the plant to investigate the problem. The individual talks to the machine operator and uses the five question technique.

Question—"Why are the parts defective?"

Answer—"The holes do not line up."

Question—"Why do the holes not line up?"

Answer—"We're using components from a new supplier."

Question—"Why are we using a new supplier?"

The question-and-answer process continues until the root cause of the problem is determined. Or in this instance the machine operator is unable to explain why there is a new supplier for the components. So the third question goes unanswered.

Fishbone or Ishikawa Diagram

So who can answer the third question? This is often the situation, that the path to the root cause involves several individuals working

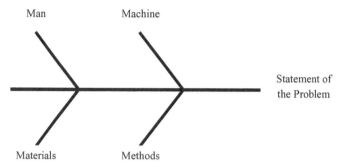

Figure 5.8. The Fishbone or Ishikawa diagram.

in different areas of responsibility. The Fishbone or Ishikawa diagram divides the possible information sources and responsibilities into several categories or classes. The four most commonly used are manpower, methods, materials, and machines (Fig. 5.8). The environment (temperature, humidity, ambient, light, etc.) is often included as a category, as is the product design. Using the above example the first step is to write down a statement of the problem: "the machine is producing defective parts." The next step is to determine which branch or branches lead to the source of the problem. In use each branch will have several sub-branches. For example, the Machine branch would include setup, proper maintenance, and so forth.

This approach to problem solving encourages collaboration and the structure also makes the problem solving approach comprehensive. It insists that all the basics of manufacturing (man, materials, machine, and methods) be identified and included in determining the root cause of the problem.

Situational Awareness

The people responsible for designing and managing the manufacturing facility (manufacturing engineering and operations) have to be aware of the regulatory requirements (local, state, and national) that affect their particular plant. Recall that in the outline

(Chapter 3) of the general regulations covering emissions, effluents, solid wastes, and runoff water there are significant variations depending on the geographical location of the plant. In some situations the growth of a community can change the circumstances that will impact the plant's operations and profitability. The point is this function (manufacturing engineering and operations) has to be involved in the community so they understand their responsibilities and the means to meet these responsibilities to remain profitable and compliant.

Situational awareness also means being aware of current and upcoming changes in product and manufacturing technologies as well as relevant regulatory obligations. The digital camera is an example of a product technology change that has practically eliminated the manufacturing of photographic film and made Polaroid's camera obsolete. Changes in style and use can diminish or eliminate a product. The Palm PDA morphed from a daily planner to a multifunctional communications device. On a less dramatic scale changes in manufacturing technology often creep into an industry. Two good examples are computer-aided design and computer-aided manufacturing. Companies that were oblivious to these technologies found themselves unable to compete or having to pass up what should have been profitable business.

The tools that a company can use to develop a situational awareness fall under the heading of "networking." A short list is:

- Have your management and technical people be active in both professional and trade organizations. This is a good source for learning about impending changes in technologies. In some cases this is where the impact of government-mandated changes are identified. An example is the changes the air conditioning manufacturers have had to make to comply with the progression of Freon-replacement refrigerants.
- Belong to a regional business group that provides information on wage and salary trends, local environmental issues, and tax law.

- Benchmark with companies in your industry and with companies that use similar workforce skills and/or technologies. This enables a company to determine if it is competitive in terms of workforce productivity. This is an excellent detector for creeping technology.
- Be proactive in participating in environmental and energy-saving programs sponsored by state agencies and utility providers. This is an excellent way to initiate change. Also, it often provides expertise and technical support that would be difficult or very expensive to find elsewhere.
- Publish accurate operational data for employees that reflect progress or lack of progress toward company goals. This is part of building trust to support a culture based on reducing waste. This also provides the basis for creating situational awareness that would be needed by company employees and associates.

PRODUCT DESIGN

Design for Assembly

In *Product Design for Manufacture and Assembly*, Boothroyd, Dewhurst, and Knight outlined how products should be designed for either manual or automated assembly. Often referred to as *Design for Assembly*, the approach explains how to ease the way products are put together and ways to eliminate errors. The authors developed three questions that a designer should ask to determine if a part is really needed:

1. Does the part have to move relative to other parts?
2. Does the part have to be made from material that is different from the other parts?
3. Does the part have a unique function that sets it apart from the other parts? An example might be a drain plug or an inspection plate.

If the answer is no to all three questions then there is a good chance the part can be eliminated. Reducing the number of parts in a product simplifies manufacturing and accordingly it will reduce costs, both material and labor. There are also precepts that should be applied to part design to make it easier to assemble, which will reduce assembly time and rework costs. Recall that our definition for cost reduction is synonymous with reducing waste. But reducing costs does not mean reducing the value of the product to the consumer.

HUMAN RESOURCES

The Life Cycle

It has been observed that *life cycles* are the way of nature. For example, a company purchases a new machine and installs it. During the first few weeks of operation there are a few start-up problems that need to be fixed—these are often referred to as working out the bugs. This is the infant phase. Once these problems are resolved the machine performs well for years. This is called normal life. However, as the years mount up the machine breaks down more often until it's decided it is time to scrap it. That's the end-of-life phase. This sequence is the life cycle of the machine.

Products have life cycles too. A product's design doesn't wear out as such but its life in the marketplace does. Newer designs, technical advancements, or style can end its life.

Jobs in a manufacturing operation also have a life cycle. This is slightly off the point of this discussion but not all jobs in a plant are "career" occupations. Most jobs should have a life cycle. When a person enters a job he or she needs some training and mentoring before entering the normal life of the job. However, the challenge, interest level, and satisfaction from doing a job can wane. This is end-of-life for the person doing this job. Some firms try to extend the life of the job by increasing pay. Although this will often convince the person to remain in the position, the benefits are limited and the additional costs in wages are waste. There has to be a progression or job renewal for the employee. When a job has reached "end-of-life" some companies move employees into different work or send them to jobs with increased responsibilities. Nevertheless employees should not be left to linger on in a job that has reached end-of-life.

A Just-in-Time Workforce

The just-in-time (*JIT*) *workforce* is not really a "tool" but more of a concept. It's a way of staffing an operation. Most managers realize that every manufacturing operation needs a core group of employees. This core group holds the knowledge, skills, history, and company culture that are vital in maintaining a profitable and compliant manufacturing operation. However, business is never static. Demand for the company's products goes up and down, causing employment levels to go up and down. During the twentieth century, companies varied employment to match demand using a seniority system; persons hired most recently would be "laid off" first. The assumption was that employees with long service would be skilled and knowledgeable of the ways the company operated. The fact that they had stayed with the company the longest would seem to imply that they were also committed to their work.

In the late 1970s cross-training manufacturing personnel became popular. The objective was twofold: create a flexible workforce and possibly extend the life of the job by creating a more interesting and challenging work experience for the individual. And, from the 1970s on, computer-based control systems of machines and processes has become commonplace in manufacturing. In general "factory work" has become more knowledge based and less physical.

Work rules, labor union contracts, and inflexible management practice can thwart a company's effort to create a flexible and a responsive workforce that can quickly react to changes in production volume. However, every company can establish a strategy to create a flexible workforce. So just what is a *flexible workforce*? It's able to respond just-in-time (JIT) to downturns and upturns in production levels and do it without generating waste. The JIT workforce retains the knowledge, skills, best practice, and culture that are essential to produce products without generating waste.

The first chapter talked about a tool and die company that had a high turnover rate with their machinists. They realized that they required individuals who had basic machinist skills but the repetitive nature caused most of their machinists to move on to more challenging jobs. Their solution was to hire machinists coming out of an apprenticeship program and graduates from machinist trade schools. This gave them a workforce that did not expect a career with the company but did value the work experience. The company did work to retain a cadre of experienced machinists who had the ability to train and mentor newly hired machinists.

Human resources identified several constraints that had to be accommodated to produce good product profitably. These constraints shaped the creation of a JIT workforce. In brief they realized they must:

- Control labor costs; their customers can buy the same tooling they produce from offshore suppliers that have comparably skilled employees who are equally productive and work for a lower hourly wage.
- Locate where there is a qualified labor source; they must be able to attract capable machinists from the local area.
- Have an elastic work force; labor requirements can change rapidly because their customers increase or cancel orders weekly, often within days of a scheduled delivery.

You already know part of their solution; they hired young machinists looking for experience but not a career. The company was able to attract machinists even though starting pay was at the low end of the wage scale. Of course there were some positive factors; they always seemed to have jobs available and the company has a good reputation in the tool and die industry, so work experience there was valued.

To deal with fluctuating demand they instituted a program to find qualified machinists who were willing to work part time. They found community college instructors, retired machinists, and machinists looking for work in the evenings. These individuals were the *"on-call workforce"* (OCW). To become part of the OCW a candidate went through the normal interview and screening process and had to successfully complete a 10-hour safety and orientation program offered at times convenient to the individual. Once they were accepted the company guaranteed them 26 hours of work but no more than 130 hours every three months.

The OCW allowed the company to handle large fluctuations in business without having to resort to layoffs or high levels of overtime. The minimum hours guaranteed to the OCW helped to maintain a company–employee relationship and provided enough financial incentive to keep them involved and value their relationship with the company.

Obviously every manufacturing facility has unique circumstances that shape employment practices that influence how the company responds to varying demand. The essential elements of any strategy, however, must provide a means to increase and decrease labor to match demand without:

- Placing the burden on the employee when it's operating management's responsibility. Examples of burden are layoffs or continuing high levels of overtime.
- Placing the burden on the customer when it's operating management's responsibility. Examples of burden are missed delivery dates, either late or early shipments, and defective product.
- Placing the burden on the company's owners when it's operating management's responsibility. Examples of burden are poor productivity compared with similar manufacturing facilities and the inability to attract business.

Developing a flexible, productive, and competent workforce is operating management's responsibility. Human resources, along with training and development, are functions within a manufacturing facility that can assist in implementing the strategy of a JIT workforce for the operating management. The results of an effective strategy will enhance a company's ability to reduce waste and be profitable.

SUMMARY

For those of you that have experience in manufacturing or are already familiar with lean manufacturing you will recognize that the techniques and concepts covered in this chapter are a long way from being a comprehensive discourse on the subject. However, the topics covered provide examples of some of the tools that can be used by the two groups and their respective functions that have the responsibility to reduce waste in a manufacturing organization.

It might be useful to summarize the key steps that have been covered. There is a sequence that may not have been immediately apparent. A first step is gathering information through sources-and-uses and benchmarking. This will tell an organization where it is at and enables it to prioritize what should be dealt with first. Establishing a company-wide culture to reduce waste is the next step. Coupled to this culture should a method of operation based on *continuous improvement*. Being able to control a manufacturing operation or process so that it is predictable is an important company-wide skill to sustain improvements. Consequently the workforce will also need good problem-solving skills.

Although it may appear that the first group—product design, marketing, and manufacturing engineering—is being overlooked, that is not the case. The real improvements in waste and resource reduction for this group can come from the PCPC that was introduced in Chapter 2. The PCPC will be developed further in the final chapter.

SELECTED BIBLIOGRAPHY

Allen, T. T. (2010). *Introduction to Engineering Statistics and Lean Sigma: Statistical Quality Control and Design of Experiments and Systems.* London: Springer-Verlag.

Boothroyd, G., Dewhurst, P., and Knight, W. (2002). *Product Design for Manufacture and Assembly*, 2nd Edition. New York: Marcel Decker.

Goldratt, E. M., and Cox, J. (1992). *The Goal.* Great Barrington, MA: North River Press.

Michalski, W., and King, D. (2003). *Six Sigma Navigator.* New York: Productivity Press.

Schonberger, R. (1986). *World Class Manufacturing, the Lessons of Simplicity Applied.* New York: Simon and Schuster.

Shingo, S. (1989). *Study of the Toyota Production System: From an Industrial Engineering Viewpoint* (A. P. Dillon, Translator). New York: Productivity Press.

Standard, C., and Davis, D. (1999). *Running Today's Factory.* Dearborn, MI: Society of Manufacturing Engineers.

Stephens, K. S., and Juran, J. M. (2004). *Juran, Quality, and a Century of Improvement.* Milwaukee, WI: ASQ Quality Press.

CHAPTER 6

THE FACILITY

INTRODUCTION

Reducing waste is not limited to the activities and processes that are operating inside the factory building. The structure itself that houses these manufacturing processes is also a generator of waste and a consumer of substantial quantities of resources. Obviously the opportunities to limit the amount of waste a facility generates and the resources it will consume depends on whether this is an existing structure or one that is in the planning stages to be built. In either case minimizing waste and resource use once again depends on the actions of the three functions described in the first chapter. Those functional groups identified a need, designed the product, and then determined how it would be made. In constructing a new plant these functional groups remain but the membership in these functions changes.

Improving Profitability Through Green Manufacturing: Creating a Profitable and Environmentally Compliant Manufacturing Facility, First Edition.
David R. Hillis and J. Barry DuVall.
© 2012 John Wiley & Sons, Inc. Published 2012 by John Wiley & Sons, Inc.

When a new manufacturing facility is needed the manufacturing technologists, production engineers, and the factory workers and other employees become the *customers* for the facility. It can be reasoned that the manufacturing technologists and the production engineers (the manufacturing/production engineering group) take on some of the responsibilities of the marketing function since they define the "manufacturing needs" for the building, but they are primarily the customers for the facility. But the others, the employees working in the building, are often overlooked for their importance as customers. According to the U.S. General Services Administration, "sustainable design seeks to reduce negative impacts on the environment, and the health and comfort of building occupants, thereby improving building performance" (see http://www.gsa.gov/portal/content/104462 [U.S. GSA]). This states why the people who work in the facility must be included as customers.

There are a variety of reasons why a company wants to construct a new facility. Usually manufacturers want a more efficient facility or they construct a new plant to better serve an important market. The *marketing function* in a company has the responsibility to detect this need. For new products this usually begins with marketing presenting a business opportunity to company management, often in collaboration with the design and the manufacturing/production engineering groups. Even with this in mind the marketing function is largely in the hands of management, where "management" refers to those individuals who are able to authorize the company or organization to commit to the construction of a building.

The second functional group is *design*. The members of this function are architects and management, with input from the manufacturing/production engineering group. Of course manufacturing/production engineering is a member of the customer group, so its input and involvement in design would be expected.

Manufacturing engineering constitutes the third functional group. In construction the manufacturing engineering is under the

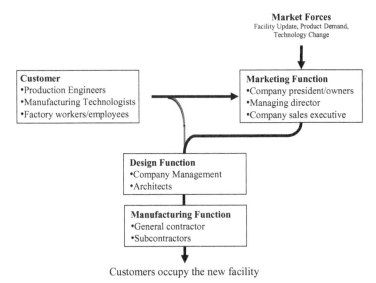

Figure 6.1. Functions involved in the construction of a new facility and their relationship to each other.

control of the general contractor (the construction company) and its subcontractors. They determine the operations and processes that will be used to construct the product, which of course is the building as situated on the grounds. Figure 6.1 shows the relationships of these groups in the three functions.

Once the building is completed it is placed in the hands of the customers. The production engineers, the manufacturing technologists, and the plant employees, with the support of management, are then responsible for placing the plant into operation. The operational costs and the environmental impact that the building will have during its operational life are largely controlled by these people.

This establishes who has what responsibility but it doesn't specifically define the building's characteristics. In general we know that it has to hold equipment and processes, have a minimal environmental impact, and provide a healthy workplace. However, there is much more to consider before the facility can be profitable and environmentally compliant. A manufacturing plant has to have the capacity for regeneration.

MAKING A BUILDING THAT CAN BE REGENERATED

"Regenerating" something means to renew, restore, and recreate it—the ability to stimulate, revive, and redevelop. Therefore a building that can be regenerated is one that is able to quickly adapt to a new use. If you recall, the first chapter gave some examples that described regeneration. A good example of a facility that can be easily regenerated is a convention center. In nearly every large city around the world there is at least one building that is capable of housing a wide variety of activities. For instance in Chicago there is the McCormick Place Convention Center, which attracts close to three million visitors each year. In September 2010, McCormick Place hosted the International Manufacturing Technology Show. The exhibition covered over 1.1 million square feet of floor space, displaying manufacturing equipment in operation. Then, two weeks later the facility hosted a completely different exhibition. Facilities such as these have lighting, services, and access that enable the building to be quickly adapted for virtually any endeavor.

For that reason a regenerative building is flexible; it is not an obstacle to change. Therefore it has a characteristic very similar to one found in lean manufacturing called SMED—*single-minute exchange of dies*. This concept implies that retooling or changing over a piece of equipment happens so quickly that there is only a minute's worth of downtime. For a manufacturing facility it means the plant floor can be easily changed over to produce an entirely different product. So instead of SMED the regenerative facility would be an *easily adaptable building* (EAB).

PLANT LOCATION

Any project that involves real estate will begin by considering site location. Transportation costs and proximity to suppliers and customers, along with the cost of living, the local tax burden, industrial infrastructure, and regional work habits, are all factors to be included in selecting the location of a new manufacturing facility. But equally important is choosing a location that minimizes the building's impact on the environment. All of these factors have to be included in the plant location analysis.

Climate costs can become a factor in site selection. The first step in estimating the magnitude of this factor is obtaining specific climate data for the facility's geographical location. Figure 6.2

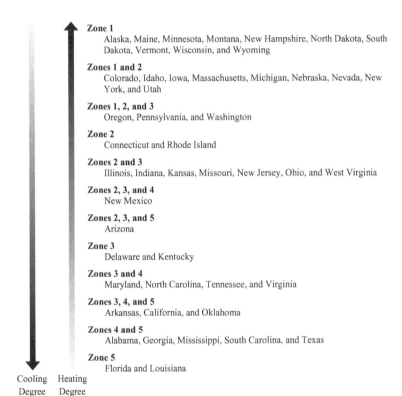

Climate Zone	Cooling Degree Days	Heating Degree Days
Zone 1	Fewer than 2000	More than 7000
Zone 2	Fewer than 2000	5500 to 7000
Zone 3	Fewer than 2000	4000 to 5499
Zone 4	Fewer than 2000	Fewer than 4000
Zone 5	2000 or more	Fewer than 4000

Figure 6.2. Annual energy demand factors for heating and cooling based on a plant's geographical location in the United States.

shows heating and cooling loads based on the geographic location of a building in the United States. This data is generated using units of measure called *heating degree day* (HDD) and *cooling degree day* (CDD). These measures are designed to reflect the demand for energy needed to heat or cool a building. If the average temperature for the day is one degree Fahrenheit above 65°F (18.33°C), it is counted as one cooling degree day. If the average daily temperature falls one degree below 65°F that is counted as one heating degree day. For example, a day with an average temperature of 80°F (26.66°C) will count as 15 cooling degree days. The annual number of HDDs is the summation of all the daily HDDs, and the annual number of CDDs is calculated in a similar way.

Figure 6.2 lists all 49 states in continental North America. Ten states fall entirely in Zone 1, two in Zone 2, two in Zone 3, and two in Zone 5. It is interesting to note that there are 33 states that fall into two or more climate zones. Comparing states based on climate zones provides some interesting contrasts. A case in point would be the comparison of Arizona and Illinois. A large portion of Arizona, which spans three zones, shares the same two climate zones as Illinois. To carry out a more detailed analysis for plant site selection will require more specific information. A good resource is the National Oceanic and Atmospheric Administration website (see http://lwf.ncdc.noaa.gov/oa/documentlibrary/hcs/hcs.html [NOAA]).

SUSTAINABLE DESIGN

So far we have two items governing the design of the building: creating a building that will not be an obstacle to change and placing it where it makes good business sense while minimizing its resource needs for heating and cooling. Next we must examine what is meant by creating a sustainable building that is compatible with its surrounding environment. A starting point is to look at the goals and objectives of sustainable building design. The description of *sustainable design* given by the U.S. Department of Commerce's International Trade Association (ITA) provides a good foundation for the concept (see http://trade.gov/sustainability):

> Over the past 30 years, the concept of sustainability has evolved to reflect perspectives of both the public and private sectors. A public policy perspective would define sustainability as the satisfaction of basic economic, social, and security needs now and in the future without undermining the natural resource base and environmental quality on which life depends. From a business perspective, the goal of sustainability is to increase long-term shareholder and social value, while decreasing industry's use of materials and reducing negative impacts on the environment. The term Sustainable Design is being used to describe a design approach that tries to reduce the negative impacts on the environment, and the health and comfort of building occupants. The basic objective of sustainable design is to reduce the use of non-renewable resources and minimize waste.

The ITA's definition covers the range of people who would have an interest in creating facilities that will not have an adverse effect on the environment now or in the future. But how would a manufacturer define the term *manufacturing facility sustainability?* As a starting point it would be a building that is able to hold up, retain its utility, and be adaptable to changes in manufacturing technology. These characteristics should also be achieved in a manner that minimizes the building's impact on the environment. This is accomplished by:

- The site meeting the needs of the manufacturing operation while being compatible with the surrounding area
- The facility being designed to retain its compatibility and compliance with the surrounding area over its lifetime
- The facility having an elegant design that would be highly effective for manufacturing and still be resource-efficient throughout its life
- Embodying the current best practices for energy efficiency using methods that must not inhibit the building's capability for regeneration

- Creating a building that will meet the environmental, health, and comfort requirements of the employees while responding to the manufacturing needs for economy, utility, and durability

It is apparent that sustainability is a complex concept. It is based on very goal-oriented architectural design, site selection and preparation, and sustainable construction methods and craftsmanship.

Sustainable design and construction practice is only part of the concept of sustainability. What has not been mentioned so far is the selection and specification of sustainable materials or stocks that would be used in construction. So the question that can be asked is "what constitutes a sustainable material?" The state of California (see www.calrecycle.ca.gov/GreenBuilding/Materials) has a useful listing of the characteristics of these materials and stocks:

- A sustainable material or stock will have low toxicity and will not contain carcinogens or contribute to indoor air pollution or health problems.
- The stocks will have minimal emissions, particularly of volatile organic compounds (VOCs) and chlorofluorocarbons.
- The material used should be either from a renewable source or from a recycled source if it is locally available and meets the other criteria for sustainable materials.
- The production of the stock should be done in an efficient manner that minimizes waste and use of resources.
- The stock should also have the potential to be recycled at the end of the building's life.
- The service life of the sustainable material should be comparable to or better than competing stocks that are not classified as sustainable.

Obviously the management and architects are largely responsible for creating the design and specifications for a sustainable building that meets these criteria. However, sustainability is an ongoing responsibility. Once the building is in operation the responsibility moves to the plant's production engineers, technologists, and employees. Certainly company management has ultimate responsibility for the facility, but sustainability is largely a tactical responsibility that requires management support. Once a manufacturing facility is constructed and in operation, the building-related wastes and resources are heavily associated with the success that the engineers, technologists, and plant employees have had in integrating the manufacturing processes with the building.

The implication therefore is that sustainability is a holistic concept starting with the building's design and continuing on through its useful life. It would seem that all of the groups involved have a massive task to create and operate a sustainable building. However, a company doesn't have to develop all the procedures, practices, and specifications since most of this has been codified in internationally recognized standards and guides.

A Sustainable Building

The American Society for Testing and Materials, which is now known as ASTM International, develops international standards for materials, products, and services used in construction, manufacturing, and transportation. One of their standards, ASTM E2432 -05, *Standard Guide for General Principles of Sustainability Relative to Buildings*, which is actually a guide rather than a series of standards, lays out the general principles of sustainability for buildings. The guide states that sustainability is based on three general principles: environmental, economic, and social. The guide covers the fundamental concepts and associated building characteristics for each of these principles and is applicable to all types of building projects. It also covers interior spaces, individual buildings, and groups of buildings along with the infrastructure systems and land use.

The guide, however, goes beyond the design and construction phase of the building. It spans the life cycle of the building. It also provides a series of options, which is why it is not a standard. A standard would recommend a specific course of action. The options therefore make it a useful reference for the three key functions responsible for the creation of a new manufacturing facility.

The guide also describes methods of decision making based on cost–benefit trade-offs and a way to specify or define building products that meet a recognized standard for sustainability. That is the purpose of ASTM E2129-05, *Standard Practice for Data Collection for Sustainability Assessment of Building Products*. Putting ASTM E2129 into a "green building" specification eliminates the vagaries that result in a material specification that refers only to "green" building products. This standard requires the contractors to purchase stocks from those suppliers that comply with a standard set of criteria defining sustainability. The ASTM E2129 standard also includes many more specific sustainability-related standards. For example:

- Test methods for determining the VOC content of coatings
- The formaldehyde content of wood
- A guide for determining organic emissions from indoor materials
- Test methods for assessing the resistance of building materials to fungus, mold, and bacterial growth
- Practices, specifications, and test methods relating to green roof design and construction
- Specifications for use of fly ash in soil stabilization and in other recycling applications
- Methods of assessing acoustical properties of building materials

You can see from this list of examples how these standards can aid in meeting the objectives for designing a "green building" that will provide a safe and healthy work environment, limit emissions, and reduce the impact that the structure will have on the environment. The philosophy of "green buildings" also encompasses the creation of a building that will not jeopardize or limit the opportunities available to future generations.

A "green building" is in tune with today's need for efficient manufacturing. If it hasn't already happened, the concept of sustainable building design will soon be recognized as a part of the discipline that manufacturers have embraced in lean manufacturing, continuous quality management, and environmental stewardship.

CONSTRUCTION SEQUENCE

The construction of a building or facility has a generalized sequence of activity that is very similar to the manufacture of a product. An example of this sequence is shown in Figure 6.3. You will notice

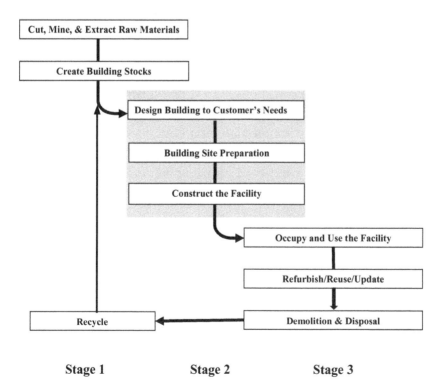

| Stage 1 | Stage 2 | Stage 3 |

Figure 6.3. The three stages in the life cycle of a building.

when you examine this sequence that it also describes the life cycle of the building.

The sequence begins by obtaining raw materials to create the stocks that will be used in the construction of the building. The construction sequence, like manufacturing, is grouped by stages. The first stage, creating stocks, includes dimensional lumber, cement, insulation, and even major components that are classified as products in manufacturing. Examples of these components are HVAC (heating, ventilation, and air conditioning) equipment, electrical distribution panels, lighting, and windows. Stage 2 is the actual construction of the building, and Stage 3 covers the occupancy and use of the building. This last stage spans the longest time and would include periods of refurbishment. Ultimately the building will be demolished and some of its materials will be recycled.

In Stage 2 the company management and architects have a significant role in determining if the building or facility will be sustainable. However, in Stage 3 that responsibility changes. The customers (production engineers, manufacturing technologists, and plant employees) join plant management and human resources as the three functions responsible for minimizing facility-related waste, resource use, and the building's ongoing environmental impact. Since the operational life of a building can last for years, decades, and possibly centuries the waste produced and resources used by the facility will be a factor that will influence the profitability of its manufacturing operation.

LIFE CYCLE AND LIFE CYCLE COST ANALYSIS

Stage 2, the planning and construction of a building, always draws a lot of attention. However, all three stages need to be considered. This is why building life cycle and life cycle cost analysis should be used to assess the impact a building will have on the environment as well as on the business.

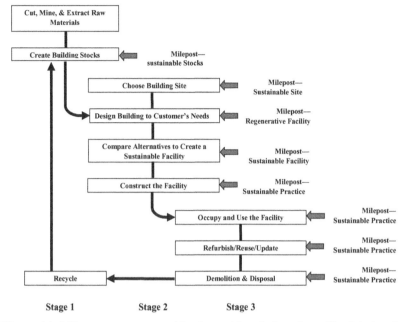

Figure 6.4. Mileposts or opportunities for making choices that will minimize the environmental impact of a building over its life.

Building life cycle analysis defines a building from its creation to its demolition and disposal. The approach as you probably suspect is similar to product life cycle analysis. Similarly, the objective for carrying out a building life cycle analysis is to identify opportunities for improvement. Figure 6.4 shows the life of a facility and the opportunities or mileposts where sustainable choices need to be made. In the last four steps of the sequence the mileposts call for sustainable practice. In these instances the practice is being profitable and environmentally compliant. If you recall, in the Corning Hickory case study (Chapter 4, Case Study 4) the ventilation system in the compressor room was modified so that the air it was exhausting was not being replaced with air-conditioned or heated plant air. The outcome was a significant reduction in energy costs. This is the result of a sustainable practice of continuous improvement being applied in the seventh step (occupy and use the facility) of the building's life cycle.

You may have noticed that Stage 2 of the life cycle sequence (Fig. 6.4) differs from the Stage 2 of the construction sequence (Fig. 6.3). The life cycle analysis is a process for planning and decision making; the construction sequence portrays activities.

Therefore site selection is a starting point in itself. Most manufacturers choose a site based on market considerations and industrial climate, which would include labor costs, tax rates, and so forth. Site selection is also subject to environmental considerations. The selection can also be influenced by energy costs, which could dictate choosing a latitude that has a mild climate or other environmental factors that would improve insurance ratings and rates.

Most of these factors or variables can be handled quantitatively on the basis of costs, both monetary and environmental. Therefore the *life cycle cost analysis* (LCCA) is an excellent method to help a company to be profitable and environmentally compliant. This approach provides an owner or company with a means to estimate and evaluate the cost of a facility, beginning with its design and continuing on until it is sold or torn down and disposed. The objective is to find the least-cost design (initial cost plus the present value of all future costs) that meets the customer's design criteria of minimizing waste, resource use, and environmental damage.

The way this is done is by evaluating alternative design approaches over time. Think about the profitable and compliant process chart (PCPC) that was introduced as a product design strategy. The PCPC is based on the concept of decision points that occur when materials and processes are chosen. Similarly the selection of each material and process used in the construction of the facility is also an opportunity to reduce waste and resource use.

As an example there is a wide variety of roofing systems available for a building located on the east coast of the United States at 36° north latitude and 100 miles from the coastline. A 50,000 square foot single-floor building could have a flat ballasted membrane roof or possibly a gently sloping living green roof. The installation costs for the two systems may differ slightly; however, the supporting structure for the competing roof systems could make one system less expensive for the initial installation. The living green roof provides some significant benefits in reducing heat loads through evaporative cooling, using retained rainwater in the soil. This rainwater retention also reduces storm water runoff, which could be a compliance issue depending on the building's location. The maintenance costs over the life of the building for each roof would also need to be calculated in the LCCA.

The example illustrates just a few of the many variables that have to be defined and the interrelationship that building systems have with each other. To carry out an effective life cycle analysis will require a good deal of data on the site location, structure configurations, design details, a facility use profile, energy costs, water and sanitary sewer costs, building materials, and process demands. As with every good marketing assessment this collection of information is essential to evaluate profitable and compliant alternatives.

This might seem to be another unbelievably complex task for a small company that is contemplating building its first manufacturing facility or for a manufacturing company that does building additions "in-house." Fortunately there are several software programs that can help quantify the costs of alternative designs, particularly in the way they affect the cost of resources to heat and cool a building. It's important to remember that the design function includes architects or technologists skilled in manufacturing facility design as well as management. The definition of management as you recall includes those individuals who have the ability to propose viable design alternatives and are decision makers.

COST ANALYSIS SOFTWARE

One of the quickest ways to get a sense of how a life cycle cost analysis is carried out is through the use of software programs tailored to do this type of analysis. A variety of programs are available and many are freeware that can be downloaded. These software programs ask for a great deal of information, so it helps to do a practice session. In the practice session you will learn what information you will need to gather, which will be the major task but well worth the time.

The eQUEST® Program

The program eQUEST is a freeware program designed to perform a cost analysis using current building practices that influence a building's energy consumption. The modeling does not require extensive experience in the "art" of building construction since the program coaches you using wizards. The two major wizards are the "building creation wizard" and the "energy efficiency measure" (EEM) wizard. There are other wizards that help complete the analysis, such as a "schematic design wizard" and a "design development wizard." There is also a graphical results display module using a DOE-2-derived building energy use simulation program. The DOE-2 is a computer program developed by the U.S. Department of Energy that predicts the hourly energy use and energy cost of a building based on weather information, a description of the building, its HVAC equipment, and the utility rate structure. Figure 6.5 shows two "screen grabs" using the eQUEST program doing an analysis of a 100,000 square foot manufacturing facility in eastern North Carolina.

The eVALUator Program

Energy Design Resources offers a variety of energy design tools and resources to help architects, engineers, and the building trades to design and construct energy-efficient commercial and industrial buildings in California (their website is http://www.energydesignresources.com/). On their website you can find energy design tools and resources as freeware for architects, engineers, and designers. One of the programs available for download is eVALUator. This program provides a life cycle analysis and a cash

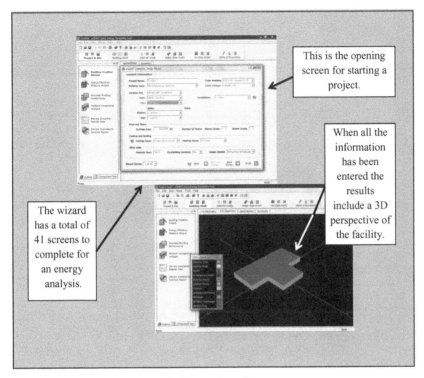

This is the opening screen for starting a project.

When all the information has been entered the results include a 3D perspective of the facility.

The wizard has a total of 41 screens to complete for an energy analysis.

Figure 6.5. "Screen grabs" from the eQuest energy use program.

flow report as part of the output. It also prompts users using a wizard approach for entering data (Fig. 6.6).

Most of the software programs such as eVALUator provide a means to compare alternative building designs over the expected life of the facility. These programs can be quite rigorous, requiring the analyst to consider a wide range of factors that can influence lifetime energy costs. This particular program examines the following:

- Financing costs
- Energy unit costs such as cost per kilowatt-hour (kWh), therm, or gallon
- Equipment replacement costs and replacement intervals
- Operation and maintenance costs

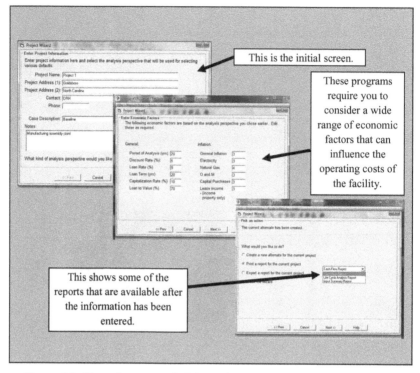

This is the initial screen.

These programs require you to consider a wide range of economic factors that can influence the operating costs of the facility.

This shows some of the reports that are available after the information has been entered.

Figure 6.6. Three "screen grabs" from the eVALUator program's wizard.

- The rate of inflation over the period of analysis
- Other benefits from the investment such as improved productivity and/or reduction in waste

BUILDING FOR ENVIRONMENTAL AND ECONOMIC SUSTAINABILITY

The *BEES* (Building for Environmental and Economic Sustainability) is web-based software that was developed by the National Institute of Standards and Technology (NIST) Engineering Laboratory to aid in the identification and selection of environmentally friendly building products. The building products in the BEES software are defined and classified according to the

ASTM standard classification for building elements known as Uniformat II, which classifies the major elements common to most buildings and related site work. The elements as defined by the Uniformat II standard classification fall into one of three levels: level 1, major group elements such as substructures for buildings; level 2, group elements such as foundations and basements; level 3, individual elements, which can be a slab on grade, a basement with walls, or a special foundation. The BEES system includes environmental and economic performance data on 230 building products. The products range from generic parking lot asphalt to brick and mortar to generic clay roofing tile on a layer of felt.

The program is intended to be used by designers, builders, and product manufacturers to create sustainable environmentally responsible structures. The evaluation method used by BEES is based on a life cycle assessment approach found in the ISO 14040 standard. This provides a comprehensive means to measure the environmental performance of building products. The economic performance of the building products is also evaluated in BEES using the ASTM standard life cycle cost method. The approach used covers the initial investment, cost of operation, maintenance-and-repair, disposal, and replacement. Figure 6.7 shows the environmental factors considered in the analysis that BEES uses to generate a score. The program is particularly useful for evaluating alternative building materials since it factors both costs and the 12 sources that affect the structure's environmental impact.

ENERGY STAR

Energy Star is a joint program created by the U.S. Environmental Protection Agency and the U.S. Department of Energy during the early 1990s to promote more energy-efficient consumer and office products. Since that time the program has been adopted by several countries, including Australia, Canada, Japan, New Zealand, Taiwan, and the European Union. For a product to carry the Energy Star label it must meet the following conditions:

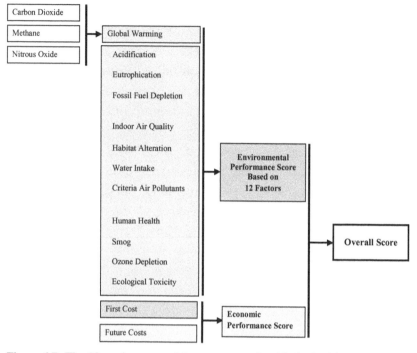

Figure 6.7. The 12 environmental impact categories (dark shade) and the economic costs that are scored for a BEES project.

- The product type must be popular enough (large market size) that it can make significant energy savings nationwide.
- Besides improved energy efficiency the product must possess the features and performance demanded by consumers.
- If the product costs more than a less-efficient competing product, the purchaser should be able to recover the additional cost from energy savings within a reasonable period of time.
- Energy efficiency should be achieved using nonproprietary technologies offered by more than one manufacturer.
- The energy consumption and performance must be able to be measured and verified through testing.

For manufacturers wanting to minimize energy use it is obviously more involved than purchasing appliances, computers, and heating and air conditioning equipment that is Energy Star rated. Manufacturing equipment and machinery is generally too specialized to meet the first condition for market size. However, the methodology used to develop Energy Star efficient products is applicable manufacturing equipment.

Energy Management Program

In the September 2002 issue of *Inside ASHE* (American Society for Healthcare Engineering), Clark A. Reed, a manager at the U.S. Environmental Protection Agency, wrote about the process of creating an energy management program (Reed, 2002). In this article he identified seven essential elements or steps to create a plan to reduce energy requirements. Figure 6.8 illustrates the essence of his process and how the steps would relate to each other.

This energy management program incorporates a process of continuous improvement. Consequently, before the program gets

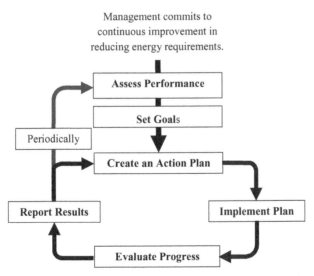

Figure 6.8. An energy management program based on the concept of continuous improvement.

started there has to be commitment. This is to dispel any notion that the program is just another perfunctory duty. Once the management group is truly committed then the first step, an assessment of current performance, can begin.

This first step will determine the magnitude of the potential energy savings to the facility's fixed and variable costs. The energy-related fixed costs are, for example, the building's basic lighting and heating-and-ventilation requirements. These are costs that will not vary due to changes in production activity. Variable costs are related to the volume of manufacturing activity. Once the energy cost basis is established goals can be set. Then the four-step continuous improvement sequence can begin, which often has the potential for being one of the most effective cost-reduction efforts a company can implement. The magnitude of these potential savings is what will underpin the organization's commitment.

Steps to Establish an Energy Management Program

The following steps outline the activities and tasks in an energy management program. However, as mentioned the first task is establishing commitment.

1. Make an assessment of the organizations energy performance by:
   - Accounting for all energy sources
   - Identifying major areas of energy consumption
   - Collecting historical data on energy consumption
   - Benchmarking current energy performance, particularly against "best practices"
2. Set performance goals.
3. Create and implement an action plan. This includes providing necessary resources.
4. Train and motivate staff as needed to carry out the plan.
5. Evaluate progress.

6. Communicate results.

7. Periodically begin again by starting at Step 1.

Some of the most detailed and difficult work begins at Step 1. This is essentially determining where, what, and why energy is used in the facility. Next is *benchmarking* the facility's energy consumption. The definition of benchmarking is comparing the performance of an item or process against a standard of performance for that item or process. As an example consider a manufacturing plant in climate zone 3 of the United States. The plant has 100,000 square feet of heated floor space. The plant manager has five years of records showing the amount of natural gas, measured in *therms* (a therm is a unit of heat equal to 100,000 British thermal units, BTUs), that was used to heat the plant each month. With this information the plant manager goes back to the utility company and asks; "how does our plant's heating costs compare to others?" The utility that supplies the natural gas can provide examples of average gas use by building size, construction type, and building use based on the number of degree days of heating. The plant manager could also go to an industry association and ask for similar data. Many industry and trade associations have a benchmarking service that can provide a standard for "typical" and "best practice" performance for many different types of equipment and processes. The essential aspect of benchmarking is finding a standard that indicates how your company's level of performance compares with others or possibly, if it's available, a benchmark of "best practice."

This concept of benchmarking is widely used in consumer goods. A common example is found on a new car "sticker." It will have the car's gas mileage and how it compares to like vehicles and there is also a best vehicle rate shown ("best practice"). Energy costs are also displayed on an Energy Guide label found on appliances. The estimated yearly operating cost for the particular applianceis shown on a line graph that portrays the yearly operating costs for similar models. Benchmarking therefore is knowing where your facility is relative to others and the current best level of performance. Once the benchmarking has been completed the next step is setting goals. This offers an opportunity to use the Pareto principle. In this case the biggest positive discrepancies in performance based on the facility's use of energy minus the benchmarked standard of performance should be used as a goal. The second application of the Pareto principle should be directed at the largest users of energy. Finally, there is another principle that should be put into use.

"Pick the low-hanging fruit" is an approach for doing the easiest tasks first. As an example consider a 600,000 square foot facility that makes lawn mowers. The electricity used in this plant is going to be substantial in comparison to the lights and the flat screen computer monitors that are always on at night and over the weekends in the 5000 square foot front office. But the wasted energy caused by this practice can be easily eliminated, although the savings compared with total plant energy use will be minimal. Nevertheless having the people in the office making a point of turning off the lights and computer screens at night is an important symbol of commitment. Examples of awareness and employee commitment are important for a program's success.

Finally, take a look at Figure 6.8 again, particularly the circular sequence that begins with "create an action plan." You will notice that this is both a beginning and an end. As noted this is the continuous improvement process. This nonstop process to improve energy efficiency is the basis for an even more comprehensive approach that can be used by manufacturers. It is another more formalized approach that can be integrated into the management of a manufacturing facility.

Certainly when a discussion centers on continuous improvement the first thing that comes to mind for most of us is the process of continuous quality improvement. That generally brings to mind the ISO quality standards. Therefore meeting standards of performance that are verified by third parties is not a new concept for manufacturers. This is the approach used in the ISO 9000 and 14000 series of standards, which are well established globally. Receiving a certification has proven to be an effective way for an organization to demonstrate that a company is subscribing to a prescribed management practice that is consistent with a standard. The ISO 50001 Energy Management System is very compatible with this approach.

INTERNATIONAL STANDARDS FOR AN ENERGY MANAGEMENT SYSTEM

The sets of standards that the International Organization for Standardization (ISO) has published covering quality (ISO 9000), the environment (ISO 14000), and energy management (ISO 50001) have formalized the management approach an organization uses in these areas. The benefits that a company realizes by adopting a management system that meets these standards can ultimately improve its competitiveness. In part this is due to its ability to perform, document, and demonstrate good performance because of the discipline the standards impose on the organization.

The ISO energy management system requires a company to establish and implement an energy policy that has specific objectives and performance targets. The definition of energy used by the standard includes "electricity, fuels, steam, heat, compressed air and other like media." The methodology used by the standard to reduce energy use is a sequence called *plan, do, check, and act* (PDCA).

The PDCA sequence establishes a continual improvement approach to energy management. The specifics for each step are first to conduct an energy use review to establish a baseline of the current usage levels. This provides the starting point for establishing objectives for reducing energy requirements. The standard also asks for *energy performance indicators* (EnPIs) to be developed. These are quantitative measures that will be used to determine the effectiveness of the energy management plan. The second step is to implement the plan. As with any continuous improvement system the third step is to monitor, measure, and record the data taken from the key indicators. This includes reporting results and comparing the performance against the expectations and objectives that have been established. The final step in the sequence is to act, determine what action is needed. If no progress has been made toward the objectives or if energy consumption has actually increased, then the action needed will be corrective action. Again recall the continuous improvement process in Figure 6.8. Creating an action plan based on results requires management commitment, problem-solving skills, technical skills, and monetary support.

If good progress has been made (verified by checking) toward the established objectives in the plan, the output of the PDCA sequence is a management review that will initiate changes to the energy policy. This will result in establishment of a new plan and new objectives, and the required commitment of resources. The cycle that is part of the philosophy of continuous improvement is reestablished. The ISO 50001 standard does not establish absolute requirements for reducing energy requirements. It relies on the commitments established in the energy policy developed by the facility. The energy standard has elements that are common to both the ISO 9000 and the ISO 14000 series of standards. Organizations that have received certification under either set of standards or both sets will see the compatibility of the approach. This means that adopting the ISO 50001 standard is not a change in management's approach. The PDCA approach for continuous improvement is one facet of the philosophy of an environmentally compliant and profitable manufacturer.

LEADERSHIP IN ENERGY AND ENVIRONMENTAL DESIGN

Leadership in Energy and Environmental Design (LEED) certification is another type of building standard concentrating on environmental performance and how a facility uses resources. It is similar to gaining certification under an ISO standard in that it is a voluntary program. A company's participation in a program of this type may be because it wants to make an environmental statement or because of a genuine concern for the impact that its business has on the environment. It is also possible that participation reduces costs. It is our opinion that for manufacturers it is desirable for all of these reasons. Lessening the impact a manufacturing plant has on the environment is a cost-reduction effort that may not have an immediate effect but will have operational benefits over the life of the plant. Also, participation shows concern and commitment to minimizing a facility's impact on the environment.

So what exactly is LEED? LEED is a rating and certification system that can be used as a benchmark in the design, construction, and operation of "green buildings." This system is intended to highlight sustainable design and construction practices that will improve a building's environmental performance. It also rates how well the structure attends to the health and well-being of its occupants. The process should begin by reviewing the building project and the LEED guidelines to determine if it would be beneficial to pursue certification.

LEED measures or rates nine areas of performance:

- Sustainable site
- Water use efficiency
- Energy and the effect on the atmosphere (such as a reduction in CO_2 emission)
- Use of materials and resources
- Indoor environmental quality
- Locations and relationship (how well does the building fit into the community)
- Awareness and education of occupants and those involved with the building
- Innovation in design (using innovation technologies that go beyond LEED requirements)
- Regional priority

Each project submitted for certification must meet minimum requirements to qualify for certification—and of course the project must comply with all applicable environmental regulations. Certification is based on a 100-point scale that is broken down into four levels of certification:

- Certified (40–49 points)
- Silver (50–59)
- Gold (60–79)
- Platinum (80 points or more)

Six points can be added to an application that shows innovation in design and an additional four points for regional priority, which brings the total points possible to 110. To gain LEED certification a facility has to submit an application documenting its compliance with the requirements of the rating system. This begins by paying the registration and certification fees. Certification is granted by the Green Building Certification Institute (GBCI), which is carried out by third-party verification of the project's compliance with the LEED requirements. There are five basic project classifications that define the eligible building types or projects for certification:

- Building design and construction
- Interior design and construction
- Building operations and maintenance
- Neighborhood development
- Home design and construction

These classifications are directed toward, but not necessarily limited to, offices, retail, institutional buildings (healthcare, libraries, schools, museums, religious, and hotels), and residential buildings. Factories and manufacturing facilities were not thought to be candidates for a LEED certification. However, the Steelcase wood furniture manufacturing plant in Caledonia, Michigan, was certified at the LEED Silver level in 2001. It was the first industrial facility to receive certification from the U.S. Green Building

Council. Ten years later 198 industrial applications had been certified for a LEED award.

SUMMARY

The tasks for designing, constructing, and using a manufacturing
facility that supports a profitable and environmentally compliant
manufacturing activity are very similar to those used to create and
produce a product. There are three functions involved in developing a new manufacturing facility and there are three involved in
its operation. Once a building is constructed and goes into operation, Stage 3 in its life cycle, many of the opportunities or mileposts
for creating a sustainable facility are in the past. However, the
responsible functions (production engineers, manufacturing technologists, plant employees, and operating management) still have
a significant opportunity to minimize the facility's energy and
resource consumption.

In the case studies in Chapter 4 we found a common thread,
an organizational and management philosophy of continuous
improvement. This approach can be summarized by the Plan, Do,
Check, and Act method described in the Energy Management ISO
50001 standard.

Data gathering and planning to create a profitable and environmentally compliant facility is an essential but burdensome activity.
Fortunately there are several computer-based resources, particularly for assessing energy use, that are available to architects and
plant engineers to assist with this task. These can be used effectively to do a "what if" analysis on existing facilities and for facilities that are being planned for construction in Stage 2 of a building's
life cycle.

The concept of sustainability is a predominate factor in the
design and construction of a manufacturing plant. The environmental impact created by the construction of the plant can be
minimized through appropriate site selection and the use of

sustainable materials. The construction industry and government have done a great deal to codify the environmental and safety aspects of materials for creating a sustainable building. Sustainability also means making a building that can be easily regenerated so that it does not limit change or adaptability of the manufacturing operation. Continuous improvement enables profitable and compliant manufacturing and profitable manufacturing sustains a business.

SELECTED BIBLIOGRAPHY

CalRecycle. Green building materials: sustainable (green) building. Available at www.calrecycle.ca.gov/GreenBuilding/Materials (accessed January 8, 2012). [Sustainable materials.]

DOE-2. eQuest energy simulation software program. Available at http://doe2.com/equest/index.html (accessed January 2, 2012).

Energy Design Resources. eVALUator. Available at http://www.energydesignresources.com/resources/software-tools/evaluator.aspx (accessed January 4, 2012).

Energy Star. Energy Star buildings. Washington, D.C.: U.S. Environmental Protection Agency and Department of Energy. Available at http://www.energystar.gov/ (accessed January 2, 2012).

IHS. ASTM E2129: standard practice for data collection for sustainability assessment of building products. Englewood, CO: IHS. Available at http://engineers.ihs.com/document/abstract/PCCMIBAAAAAAAAAA (accessed January 4, 2012).

ISO. ISO launches ISO 50001 energy management standard 50001. Geneva: International Organization for Standardization. Available at http://www.iso.org/iso/pressrelease.htm?refid=Ref1434 (accessed January 4, 2012).

ISO. ISO management system standard for energy. Geneva: International Organization for Standardization. Available at http://www.iso.org/iso/hot_topics/hot_topics_energy/energy_management_system_standard.htm (accessed January 4, 2012).

ITA. Sustainable design. Washington, D.C.: International Trade Administration. Available at http://trade.gov/sustainability (accessed January 4, 2012).

LBNL. Climate zones. Washington, D.C.: U.S. Department of Energy's Lawrence Berkeley National Laboratory. Available at http://

energyiq.lbl.gov/EnergyIQ/tooltips/CBClimateMap.html?width= 650&height=700 (accessed January 8, 2012). [Map.]

NIBS. Building life cycle cost analysis. Washington, D.C.: National Institute of Building Sciences. Available at http://www.wbdg.org/resources/ lcca.php (accessed January 2, 2012).

NOAA. Heating and cooling degree day data. Washington, DC: National Oceanic and Atmospheric Administration. Available at http://lwf. ncdc.noaa.gov/oa/documentlibrary/hcs/hcs.html (accessed January 31, 2012). [Climate data.]

Quality Digest. ISO 50001 energy management standard impacts the bottom line. Available at http://www.qualitydigest.com/inside/quality-insider-news/iso-50001-energy-management-standard-impacts-bottom-line.html (accessed January 4, 2012).

Reed, C. (2002). Creating a high performance energy management strategy. *Inside ASHE.* September 2002. Available at http://www.energystar. gov/ia/business/healthcare/ashe_sep_2002.pdf (accessed January 4, 2012).

U.S. DoE. Federal Energy Management Program, Building Life-Cycle Cost (BLCC) Programs. Washington, D.C.: U.S. Department of Energy, Office of Energy Efficiency and Renewable Energy. Available at http://www1.eere.energy.gov/femp/information/download_blcc.html# blcc (accessed January 4, 2012).

U.S. EPA. Sustainability. Available at http://www.epa.gov/sustainability/ (accessed January 4, 2012).

U.S. EPA. What is sustainability? Washington, D.C.: Environmental Protection Agency. Available at http://www.epa.gov/sustainability/ basicinfo.htm#sustainability (accessed January 4, 2012).

U.S. GSA. Sustainable design program. Washington, D.C.: General Services Administration. Available at http://www.gsa.gov/portal/content/ 104462 (accessed January 4, 2012).

CHAPTER 7

APPLYING THE PROFITABLE AND COMPLIANT PROCESS CHART

INTRODUCTION

Back in Chapter 2 we introduced the concept of the *profitable and compliant process chart* (PCPC). Our discussion included a scenario of the "thinking" and "doing" processes that Sally went through to make a product, tomato stakes. She worked through several *decision points* that led to a "product design" and a series of manufacturing processes to produce the stakes. This was documented in an elementary version of a PCPC. As you probably recognize, the PCPC is based on a process flow chart. Also much of our dialogue so far in this book has been on the structure of manufacturing, material stocks, manufacturing processes, data collection, and improvement methods. What we need to do now is to bring all these concepts together, showing how they come into play when manufacturing a more realistic product. A good starting

Improving Profitability Through Green Manufacturing: Creating a Profitable and Environmentally Compliant Manufacturing Facility, First Edition.
David R. Hillis and J. Barry DuVall.
© 2012 John Wiley & Sons, Inc. Published 2012 by John Wiley & Sons, Inc.

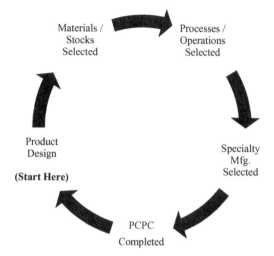

Figure 7.1. PCPC cycle of continuous improvement.

point might be to study Figure 7.1, the PCPC cycle of continuous improvement.

PCPC WORKSHEETS

The PCPC cycle starts with the product design. As part of the design procedure the material stocks are selected using the *material selection worksheet*, presented in Figure 7.2.

A material worksheet is completed for each different part, component, or subassembly of the product. This is where the design team needs to make careful decisions about the best material (stock) for the job and consider the wastes that will be created if this material is selected. The competing materials are noted on individual material selection worksheets. The selected stock would have the lowest worksheet "value." The next step after all materials are selected is to choose the manufacturing processes. This is where the manufacturing engineers get really involved. At every step in the process flow sequence a process worksheet would be completed. Process alternatives are evaluated in much the same

The Material Selection Worksheet

Product Name:
Part Number:
Component ID

Worksheet Number:
Revision Number/Date:

Competing Material IDs	Sustainability			Fitness for Use	Cost	Manufacturability	Total Value
	Creation of Stock	Manufacturing Waste	Distribution, Service, and Disposal				
MS a: *							
MS b:							
MS c:							

* MS stands for Material Selection and the alpha denotes the candidate ID being evaluated.

Scoring Key

Sustainability

Creation of Material Stock

Material resource has minimal environmental impact.	0
Nonrenewable resource has minimal environmental impact or a renewable resource with some environmental impact.	1
Material resource has a high environmental impact and/or is toxic.	2

Manufacturing Wastes

Generates little waste and requires limited resources.	0
Waste can be recycled and resource requirements are typical of alternative materials.	1
Generates solid, liquid, or gaseous waste that requires control and/or resource requirements are high compared to alternative materials.	2

Distribution, Service, Disposal

Resource requirements and waste potential are limited.	0
Resource requirements are typical and waste can be recycled—service life is comparable to other materials.	1
Resource requirements are high and/or waste requires special handling.	2

Fitness for Use

Excellent for the product's use.	0
Acceptable for the product's use.	1
Marginally suitable for the product's use.	2

Cost

Cost-effective	0
Competitive	1
Higher than acceptable alternatives	2

Manufacturability

Preferred	0
Acceptable	1
Difficult	2

Figure 7.2. Material selection worksheet.

183

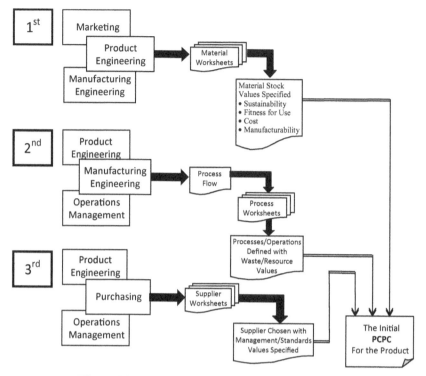

Figure 7.3. Plant floor data collection for the PCPC.

way design engineering evaluated alternative materials before choosing a specific stock. The process with the lowest value is one that would be selected.

Before we go much further let's take a look at the activities that occur on the plant floor to gather information for use in the PCPC (refer to Fig. 7.3).

The sequence of events in Figure 7.3 illustrates how three types of worksheets: material selection worksheet, process worksheet, and supplier worksheet are used to collect critical input data for the PCPC. These data are shown as "*load values*" (total from ratings addressing selection criteria) when producing the product. The total load value on the PCPC reflects the potential for waste that the product will have over its life cycle. To gain an understanding of how this impact is determined you will need to look closer at the Scoring Key in the material selection worksheet (Fig. 7.2).

Once the PCPC is completed it becomes the benchmark for that product. From that point on the company's task is to find ways to improve—to lower the score on the PCPC. This requires all the groups or functions within the company, not just the product design group, to seek ways to minimize the waste and resources used to produce the product.

This is not the only effort or even the primary effort to reduce waste. The PCPC is a system that relies on a manufacturing environment that fully utilizes the concepts of lean manufacturing. If you recall the case studies there were scores of examples where waste in time, materials, and resources were reduced. And these were improvements that did not directly impact the product. They did, however, reduce the cost of production and the impact the manufacturing operation had on the environment.

You recall that stock can be classified according to *major material families* (MMFs): woods, plastics, metals, ceramics, and composites. For every manufacturing process and material combination there is potential for waste to be generated and resources to be expended. Some combinations are better than others. Our goal is to apply the best combination. As mentioned before, the selection of the material stock(s) for the product is necessary before advancing to the next step (step 2) in Figure 7.3. This is where the manufacturing processes are selected. For most manufacturers a production or manufacturing engineer will have to establish a "routing" or "production flow chart" that will list the steps or operations needed to manufacture the product. This is essentially a process flow chart. This chart identifies all the processes or operations that will be examined using *process selection worksheets*. We need to emphasize that each process requires its own process worksheet. The worksheet has two purposes. First it lists and provides a value for the potential for waste and its demand for resources. Its second use is to provide a method for comparing alternative processes (if there are any) that would be more effective. Study Figure 7.4, the process selection worksheet.

The Process Selection Worksheet

Worksheet Number:
Revision Number/Date:

Product Name:
Part Number:
Component ID:
List ancillary materials used in the process/operation for each PC ID:

Material Name:	Description:	Material Identification:

Process /Operation Sequence Number:

Competing Processes or Operations	Solid Waste/ Unit	Liquid Waste/Unit	Gaseous Waste/Unit	Toxic or Hazardous Materials	Water Waste/Unit	Energy Waste/Unit	Total Value
PC a*							
PC b							
PC c							

* PC stands for Process Candidate and the alpha denotes the ID for the process being evaluated.

Scoring Key

Solid, Liquid, and Gaseous Waste

0	Process generates no waste. No toxic materials or elements are involved.
1	Process generates some waste which can be recycled or easily abated. No toxic materials or elements are involved.
2	Process generates waste which must be disposed of or is not easily abated.

Toxic or Hazardous Materials

0	The process does not use toxic or hazardous materials.
1	The process uses toxic or hazardous materials but effective control is possible and there is no release to the environment.
2	People are exposed to the materials and release to the environment occurs or is likely.

Water

0	The process does not use or require water.
1	Some water is required and recycling can be accomplished.
2	Water is required and recycling is not possible and/or treatment is required for disposal.

Energy

0	The process does not use or require energy.
1	Some energy is required but "best practice" is in use to limit energy use.
2	Energy is required and little has been done to limit the amount required.

Figure 7.4. Process selection worksheet template.

In practice, product and manufacturing engineering (steps 1, 2, and 3 in the PCPC cycle of continuous improvement) are primarily responsible for developing the first version of the process flow chart to make the product. As we mentioned before, the process flow chart lists the steps or *operations* for completing a product, process, or subassembly. The term "operations" refers to any action steps that occur in the manufacture of the product. They may be processes or they can also be tasks such as material handling, inspection, and delays (temporary storage). If a task specifies a particular manufacturing process that belongs to one of the five *basic process classifications* (BPCs): forming, separating, joining, conditioning, and finishing, it requires a process selection worksheet to be prepared. In each of these classifications there are many different manufacturing processes that are potential candidates for use with the stock that has been selected. This may provide a practical alternative process to be included on the PCPC. The goal of the worksheet is to identify the best process for a stock at that step in the manufacturing sequence.

What is important to remember is that the PCPC model is created using data from a set of process worksheets. The advantage of this approach is the best processes are selected based on criteria such as manufacturability and the minimum wastes required to produce the product, not on what has always been done or might be most convenient. Applying the process worksheet to identify the best process in the sequence of operations is much like the approach taken by design engineering when they evaluated alternative materials before choosing a specific stock. The process with the lowest value is the one that would be selected.

Today it is unlikely that a company that designs and assembles a product will manufacture every part, component, or subassembly that goes into the final product. Often it is more cost effective to outsource some elements to a specialty manufacturer (step 3). The manufacturing firm designing and producing parts of the product in-house will need to make careful decisions before outsourcing. Consequently the manufacturer that is chosen to supply the component should be evaluated in a manner similar to the

Supplier Selection Worksheet
Worksheet Number:
Revision Number/Date:

Product Name:
Part Number:

Names of Potential Candidate Suppliers (CS) ID:
CSa:
CSb:

Supplier ID	Management	Industry Standards	Individual Standards	Response to Environmental Regulations	Total Score
CSa					
CSb					

* CS stands for Candidate Supplier and the alpha denotes the ID for the company being evaluated

Scoring Key

Management	
0	The company has ISO certification for Quality, Environmental, and Energy Management systems.
1	The company has a Quality management system certification and is implementing Environmental and Energy management systems.
2	The company is not an active member of industry, trade, and professional organizations and is unable to document that it follows industry standards and best practices.

Industry Standards	
0	The company is an active member in industry, trade, and professional organizations and can document that it follows industry standards and best practices.
1	The company can document that it follows industry standards and best practices.
2	The company is not an active member of industry, trade, and professional organizations and is unable to document that it follows industry standards and best practices.

Individual Standards	
0	The company actively seeks to train and/or facilitate individuals to be certified for their specific skill or profession. Where certification programs are not available the company has a defined and ongoing training program.
1	The company has a defined and ongoing training program for specific skills and professions.
2	The company has limited or no training programs in place.

Company's Response to Environmental Regulations	
0	Company has been recognized by a regulatory agency as going beyond the minimum to meet environmental regulatory requirements.
1	Company has not been cited for violating environmental regulations.
2	Company has been cited for exceeding limits.

Figure 7.5. Supplier selection worksheet template.

selection of material stocks and processes. Here the company that designed the product will use the *supplier selection worksheet* shown in Figure 7.5 to carry out this assessment. You'll notice that the part or component cost criteria are not included on the supplier selection worksheet. We expect that meeting a cost target is a precondition to being considered as a supplier. Take a moment to study Figure 7.5.

Once the PCPC is completed it becomes a benchmark for that product. It is a living tool that the company can use from that point on to make continuous improvement—by putting into practice changes that will lower the total value on the PCPC. Using the PCPC requires cross-disciplinary thinking from all of the groups or functions in the company, not just in product design, to minimize the waste and resources used to manufacture the product.

USING THE DATA COLLECTION WORKSHEETS

The following example will illustrate how the three different types of worksheets—the material selection worksheets, process selection worksheets, and supplier selection worksheets—will be used to obtain and classify information for use on the PCPC. The product for our first simulation is an air rifle pellet. Refer to Figure 7.6, which shows the Diabolo air rifle pellet.

Step 1: Material Selection

A project design team was established with team members from marketing, product engineering, and manufacturing engineering. The job to be done by this team was to create the product specifications and specify the material stocks for production of the air rifle pellet. The team saw opportunity to break from the old way of producing the pellet and decided to provide an improved solution using nontraditional materials and processes. Three materials were selected as candidates for producing the air rifle pellet and compared on the materials selection worksheet. These materials

Figure 7.6. Diabolo air rifle pellet.

are (1) lead-free performance ballistic alloy, (2) tin/copper alloy, and (3) biodegradable thermoplastic resin. Figure 7.7, presents the material selection worksheet for the Diabolo air rifle pellet.

Data from the material selection worksheet indicates that of the three materials considered for the product MS 1, lead-free performance ballistic alloy, had the lowest score or load value with 4 points. MS 2, tin/copper, had 9 points, and MS 3, biodegradable thermoplastic, had a score of 6 points. For this product the major factor influencing material choice was function, or fitness for use. Based on research from the marketing team, biodegradable thermoplastic pellets did not meet the fitness for use requirements demanded by professional shooters. Therefore, the performance ballistic alloy was the material of choice.

Step 2: Process Identification

Once the material stock for the product has been selected it is time to advance to Step 2. In this step product engineering, manufacturing engineering, and operations management select the manufacturing processes. For most manufacturers a production or manufacturing engineer will have to establish a "routing" or "process flow chart" that will list the steps or operations needed to manufacture the product. (See Fig. 7.8.)

The Material Selection Worksheet

Product Name: **Air Rifle Pellet**
Part Number: **1**
Component ID Pellet Body

Worksheet Number: 1
Revision Number/Date: 1-11-15

Competing Material IDs	Sustainability			Fitness for Use	Cost	Manufacturability	Total Value
	Creation of Stock	Manufacturing Waste	Distribution, Service, and Disposal				
MSa: Lead-free Performance Ballistic Alloy	1	1	1	0	0	1	4
MSb: Tin/Copper Alloy	2	1	2	1	2	1	9
MSc: Biodegradable thermoplastic resin	2	1	1	2	0	0	6

* MS stands for Material Selection and the alpha denotes the candidate ID being evaluated.

Scoring Key

Sustainability

Creation of Material Stock

Material resource has minimal environmental impact.	0
Nonrenewable resource has minimal environmental impact or a renewable resource with some environmental impact.	1
Material resource has a high environmental impact and/or is toxic.	2

Manufacturing Wastes

Generates little waste and requires limited resources.	0
Waste can be recycled and resource requirements are typical of alternative materials.	1
Generates solid, liquid, or gaseous waste that requires control and/or resource requirements are high compared to alternative materials.	2

Distribution, Service, Disposal

Resource requirements and waste potential are limited.	0
Resource requirements are typical and waste can be recycled—service life is comparable to other materials.	1
Resource requirements are high and/or waste requires special handling.	2

Fitness for Use

Excellent for the product's use.	0
Acceptable for the product's use.	1
Marginally suitable for the product's use.	2

Cost

Cost-effective	0
Competitive	1
Higher than acceptable alternatives.	2

Manufacturability

Preferred	0
Acceptable	1
Difficult	2

Figure 7.7. Material selection worksheet for Diabolo air rifle pellet.

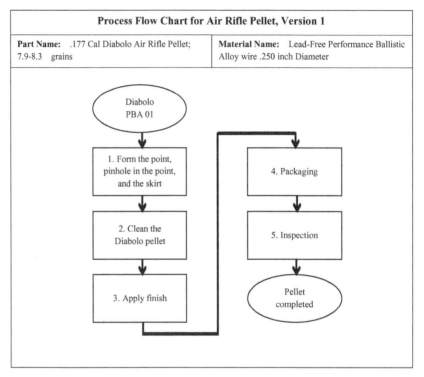

Figure 7.8. Basic process flow chart showing the sequence of manufacturing.

This chart identifies all the processes or operations that will be examined and refined using process worksheets. When this step has been completed the team will create the first version of the process flow chart for production of the air rifle pellet (see Fig. 7.9). Both of these charts are based on the material selected in Step 1, lead-free performance ballistic alloy.

We need to emphasize that each process on the process flow chart requires its own review and analysis by assigning load values from the process worksheet Scoring Key. The process worksheet has two purposes. First it lists and provides a value for the potential for waste and resource demand. Its second use is to provide a method for comparing alternative processes (if there are any) that would be more effective. Study the process selection worksheet for production of the air rifle pellets. See Figure 7.10.

Process Flow Chart for Air Rifle Pellet, Version 1

Part #: P01	Part Name: .177 Cal Diabolo Air Rifle Pellet; 7.9-8.3 grains	Material ID: PBA 01	Material Name: Lead-Free Performance Ballistic Alloy wire .250 inch Diameter
Operations	**Process Action**	**Description**	**Manufacturing Process**
1	Forming of the point, pinhole in point, and skirt.	The stock is .250″ lead-free Performance Ballistic Alloy wire that is pressed at room temperature in a swaging machine using a four-part die set. Two parts of the die clamp together to form the dome-shaped point. The third part of the die called the inner punch creates a pinhole in the tip that allows excess lead to be extruded. This is done to achieve uniform weight and also minimizes the need for removal of excess metal when finishing. The fourth part of the die is a punch that presses lead from the back end of the die to form the skirt on the pellet.	High Performance Cold Heading with an automatic swaging machine and four-part die
2	Cleaning of the Diabolo pellet.	This process uses high frequency sound waves to agitate an aqueous/organic cleaning medium to remove contaminants such as dust, dirt, oil, pigments, grease, polishing compounds, and mold release agents.	Ultrasonic Cleaning
3	Finishing	Application of coating to provide lubricity and corrosion protection	E-Coating
4	Packaging	Pellets are dispensed into a metal tin.	Auto Filling of Tin
5	Inspection	Filled tin is conveyed to the final inspection where the can is weighed and capped.	Inspection and Capping

Figure 7.9. Process flow chart for Diabolo air rifle pellet.

The Process Selection Worksheet

Product Name: .177 Cal., Diabolo Air Rifle Pellet
Part Number: P01
Component ID:

Worksheet Number:
Revision Number/Date: 11-1-15
List ancillary materials used in the process/operation for each PC ID:
PC1: aqueous/organic cleaning medium
PC2: aqueous/organic cleaning medium
PC3: ground-up corn cobs
PC4: aqueous/organic cleaning medium

Material Name: MS a: Lead-free Performance Ballistic Alloy	Description: Cleaning through removal of contaminants including dust, dirt, oil, grease, polishing compounds, and mold release agents		Material Identification: MS 1A

Process /Operation Sequence Number: PO 2

Competing Processes or Operations	Solid Waste/Unit	Liquid Waste/Unit	Gaseous Waste/Unit	Toxic or Hazardous Materials	Water Waste/Unit	Energy Waste/Unit	Total Value
PC1a:*Ultrasonic cleaning	0	2	0	1	0	0	3
PC2b: Degreasing	0	2	0	1	1	1	5
PC3c: Tumbling	3	0	0	0	0	1	4
PC4d: Washing	0	2	0	1	2	2	7

* PC stands for Process Candidate and the alpha denotes the ID for the process being evaluated.

Scoring Key

Solid, Liquid, and Gaseous Waste

Process generates no waste. No toxic materials or elements are involved.	0
Process generates some waste which can be recycled or easily abated. No toxic materials or elements are involved.	1
Process generates waste which must be disposed of or is not easily abated.	2

Toxic or Hazardous Materials

The process does not use or toxic or hazardous materials.	0
The process uses toxic or hazardous materials but effective control is possible and there is no release to the environment.	1
People are exposed to the materials and release to the environment occurs or is likely.	2

Water

The process does not use or require water.	0
Some water is required and recycling can be accomplished.	1
Water is required and recycling is not possible and/or treatment is required for disposal.	2

Energy

The process does not use or require energy.	0
Some energy is required but "best practice" is in use to limit energy use.	1
Energy is required and little has been done to limit the amount required.	2

Figure 7.10. Process selection worksheet 1 for air rifle pellet.

The worksheets shown in Figures 7.7 and 7.10 illustrate how data can be collected and used to select the best material or process. In Figure 7.10 four processes were under consideration as candidates to perform the cleaning process. Ultrasonic cleaning with a score of 3 points was the preferred process.

Step 3: Outsourcing Manufacturing Processes

Step 3 of the process flow chart lists a finishing operation that is conducted using the E-coating process. After the lead-free performance ballistic alloy was selected, the product engineering and marketing people determined that the corrosion protection required for the pellet also gave them an opportunity to improve the pellet's performance by using a protective coating material that also provides lubricity. However, their company doesn't have the facilities in-house to carry out this process. Therefore engineering forwarded a design specification to the purchasing department asking them to identify companies that can provide this type of coating and determine if they would be prepared to submit a quotation. Part of the process for requesting a quotation from a supplier is gathering information on the company and on the way the company operates. In this case several companies responded and now a supplier has to be selected.

The evaluation process involves an examination of the companies, using a supplier selection worksheet (Fig. 7.5). Data was collected from the two potential suppliers: AEAO Coating and Covering and Stokowski E-Coatings Incorporated. The worksheet for Stokowski E-Coatings is shown in Figure 7.11. Content from the supplier candidate selection worksheets will be used by purchasing to determine who can bid on coating the air rifle pellets. In this case Stokowski E-Coatings was selected as the preferred supplier.

Summary

In this example we selected a product for our air rifle pellet simulation that had only five processes in the process flow and only

The Supplier Selection Worksheet

Product Name: .177 Cal., Diabolo Air Rifle Pellet
Part Number: P05

Worksheet Number: CWS 1
Revision Number/Date: 1-11-15

Names of Candidate Supplier (CP) ID:
CSa: Stokowski E-Coatings
CSb: AEAO Coating and Covering

Scoring Key

Supplier ID	Management	Industry Standards	Individual Standards	Response to Environmental Regulations	Total Score
CPa	1	0	0	0	1
CPb	1	1	1	1	4

Management

0	The company has ISO certification for Quality, Environmental, and Energy Management systems.
1	The company has a Quality management system certification and is implementing Environmental and Energy management systems.
2	The company is not an active member of industry, trade, and professional organizations and is unable to document that it follows industry standards and best practices.

Industry Standards

0	The company is an active member in industry, trade, and professional organizations and can document that it follows industry standards and best practices.
1	The company can document that it follows industry standards and best practices.
2	The company is not an active member of industry, trade, and professional organizations and is unable to document that it follows industry standards and best practices.

Individual Standards

0	The company actively seeks to train and/or facilitate individuals to be certified for their specific skill or profession. Where certification programs are not available the company has a defined and ongoing training program.
1	The company has a defined and ongoing training program for specific skills and professions.
2	The company has limited or no training programs in place.

Company's Response to Environmental Regulations

0	Company has been recognized by a regulatory agency as going beyond the minimum to meet environmental regulatory requirements.
1	Company has not been cited for violating environmental regulations.
2	Company has been cited for exceeding limits.

Figure 7.11. Supplier selection worksheet (Process 5, E-coating).

one material other than the coating provided by an outside supplier. In the second step, where the product was cleaned, there were ancillary (secondary) materials to be considered. These are materials needed to manufacture the product but do not become part of the product. In the process selection worksheet the ancillary materials were the primary sources for liquid or solid waste, which is often the case.

Undoubtedly you have already recognized that the worksheets will need to be modified or expanded to handle the products or components your company is manufacturing. Please remember that each process and new material creates a decision point where a judgment has to be made that will determine the amount of waste generated. One other aspect that should be included in your PCPC deals with events other than manufacturing processes. Examples are delays, inspections, storage, assembly, packaging, and the resources the facility uses to support these processes that create the final product. These too are decision points where judgments must be made. The total score provided by the PCPC serves as a baseline that the firm can use in evaluating its progress toward improving the manufacturing operation.

To help illustrate some of the ways the PCPC can be used we have included a few more examples. The first three applications of the PCPC system were completed by individuals working in industry. The Scoring Key categories and values used in these examples are the same as those used with the Diabolo pellet. The fourth application will show how you can use the PCPC in an entirely different way. This application involves a capital improvement to a manufacturing facility. The Scoring Key is gone so now the evaluation of waste and energy is expressed in terms of dollars.

INDUSTRIAL APPLICATIONS OF THE PCPC

The individuals that developed these applications of the PCPC were in a graduate program offered by the Department of

Technology Systems at East Carolina University. Their applications of the PCPC provided some innovative uses of the technique
in a variety of different types of manufacturing companies. The
company names and data presented in the following cases have
been modified to preserve confidentiality. We should point out
that it is the approach that was taken by the individual that is of
value and not the specific conclusion of the study.

Application 1: Avionic Systems, Incorporated

Type of Manufacturer: Aerospace gas turbine engine
manufacturer

Objective: Document the rotor assembly operation and establish a baseline for a waste and resource minimization program.
Refer to Table 7.1.

Comment   This is an application from a Stage 2, Group 2 manufacturer. You will recall that Stage 2 manufacturers produce hard
good consumer products. In this case the product is a rotor assembly used with a gas turbine engine. As a Group 2 manufacturer
they assemble the components for the company that designed and
developed the product. A significant decision point for a Stage 2,
Group 2 manufacturer is in the selection, use, and control of secondary stocks. Being able to visualize the operations and processes will help in identifying these opportunities. For this reason
the investigator incorporated photos in the description of the
assembly operations.

In terms of the data conveyed with this partial view of the
PCPC it can be seen that operation 6, cleaning of the balancing
equipment, offers the most opportunity for improvement. The
numbers on this PCPC are low because this analysis involves
assembly operations and few manufacturing processes. This is not
unusual because the waste potential for assembly operations is
typically very low.

The use of photos was a useful enhancement and raised an
interesting point. This technique for waste reduction is not tied to

TABLE 7.1. PCPC for Avionic Systems, Inc.

Profitable and Compliant Process Chart (Partial)

Part #: ARSC GEO		Product Name: Rotor Assembly	Material ID: Steel and Alloys	Material Name:			
Operations	Mfg. Process	Description		Material Waste	Resources Used	Energy Used	Total
1	Prebalance each component			0	0	1	1
2	Install forward shaft adapter	The forward shaft adapter is positioned on the shaft. This is the surface that the rotor will be positioned against during balancing.		0	0	0	0
3	Assemble rotor component to tie bolt	The final slave bearing is installed on the opposite end of the assembly to aid in balancing.		0	0	0	0
4	Tighten tie bolt to specs	The tie bolt cap nut is torqued to spec while the components are compressed under load.		0	0	1	1
5	Install slave bearings	Final slave bearing is installed on the opposite end.		0	0	0	0
6	Clean balancing equipment	Balance machine rollers are cleaned.		1	1	1	3
7	Open electronic work instructions	Work instructions for balancing this model are referenced.		0	0	0	0
8	Retrieve saved setup data	The balancing program is accessed.		0	0	1	1
9	Adjust support heights	Rotor machine guides are set to correct heights.		0	0	0	0
10	Adjust support distances	Rotor machine guides are set to correct location.		0	0	0	0
			Totals	1	1	4	6

filling out a "paper" form. The PCPC can be a computer spreadsheet arranged in a similar manner to its paper counterpart. Then a column of cells can be formatted to hold photos of the process or possibly insert a link to the company's Intranet that holds a video of the process in operation. This can be useful when individuals from the various functions are working together. The PCPC can be projected onto a screen and bring the plant operations right into the conference room.

Application 2: American Automotive Corporation

Type of Manufacturer: Automotive industry component manufacturer

Objective: Document improvement possible by changing a material stock. Refer to Table 7.2.

Comment   This company is a Stage 2 Group 1 manufacturer that has acted on a decision point; selecting a material stock. This case compared an existing primary stock, a rubber elastomer, against a proposed thermoplastic elastomer. In this sequence of processes the current material generated a total score of 33 points and the proposed stock scored only 12 points. This resulted in a reduction of 21 points. This is a good example of continuous improvement involving the group responsible for product design. Much of the discussion on continuous improvement has focused on the operations group that manufactures the product. However, as this application illustrates the "design function" has an ongoing responsibility to improve the product by changing material stocks to minimize waste.

Another observation can be made. This application used the ordinal Scoring Key. This key is very basic but in this instance it was able to show a sizable difference between the two materials being evaluated. This Scoring Key does have utility and because of its ease of use can be a good starting point in the adoption of the PCPC.

TABLE 7.2. PCPC for American Automotive Corporation

Profitable and Compliant Process Chart

Part #: AAC Cam Mirror		Product Name: Side View Mirror Patch Seal	Material ID: AAC Cam-14	Material Name: TPE Thermoplastic Vulcanite		
Operations	Mfg. Process	Description	Material Waste	Resources Used	Energy Used	Total
Rubber Elastomer Material (Before Improvement)						
1	Feeding extruders	Two different rubber strips are fed into the extruders.	2	0	2	4
2	Melting and mixing	Raw material is melted, mixed, and extruded through a die.	2	0	2	4
3	Heating	Heating to 1000°F in natural gas operated super jet oven	0	0	2	2
4	Microwave	Microwaved	0	0	2	2
5	Heating	Pushed through a series of six natural gas operated ovens at 650°F	0	0	2	2
6	Cooling	Cooled down by air and water to 300°F	0	1	2	3
7	Treating	Prepared for coating in plasma tree machine	2	0	2	4
8	Coating	Covered with coating material in paint bath	2	0	2	4
9	Heating	Heated in natural gas operated conventional ovens (650°F)	0	0	2	2
10	Cooling	Cooled down by air and water	0	0	2	2
11	Finishing		2	1	2	4
		Total	10	1	22	33
Thermoplastic Elastomer (After Improvement)						
1	Feeding	Three types of granulated TPE materials (different hardness) are fed into three different extruders.	1	0	2	3
2	Melting and mixing	Granules are melted, mixed together, and extruded through a die.	0	0	2	2
3	Cooling	Cooled down (water)	1	1	2	4
	Finishing		1	0	2	3
		Total	3	1	8	12

Application 3: NAVAC Logistics Center

Type of Manufacturer: Helicopter remanufacturing

Objective: Evaluate and compare the waste and resources required for removing paint from a helicopter airframe, using the PCPC. Refer to Table 7.3.

Comment   NAVAC recognized that their plastic blast medium that was used in the stripping process was too aggressive and was damaging the floorboards of the H-65 aircraft. Many floorboards required rework and some were damaged beyond repair because of the abrasive impact caused by the plastic medium. Several different types of blast media were evaluated using the PCPC. In this application, the PCPC was used to compare the performance of secondary materials. You may recall that a secondary material is essential to manufacturing but it is not a part of the product. The paint that is applied later to the airframe is a primary material or stock.

One of the alternatives considered as a replacement was a blast medium made from cornstarch, eStrip GPX® (Alternative 1, operation 3 in the PCPC shown in Table 7.3). This medium was effective in removing epoxy/polyurethane paints, rain erosion resistant coatings, and radar-absorbing materials. The value in the Material Waste column was a "1," which is the ordinal value defined in the Scoring Key on the process selection worksheet. In this particular application of the PCPC a more discriminating waste measure would helpful.

Using "cost" as a criterion for assessment would help in differentiating the potential for waste reduction. The cost components for this operation would be the disposal cost of the paint and blast media along with the cost to repair or replace the damaged floorboards caused by a particular blast medium. However, changing metrics to a cost measure in this operation presents a problem. If one classification, material waste for instance, is changed to a cost measure, then all other categories must be changed to that measure. If the unit of measure is not

TABLE 7.3. PCPC for NAVAC Helicopter Cleaning and Repainting

Profitable and Compliant Process Chart (Partial)

Part #: H-65-Helicopter Cleaning and Repainting	Product Name: Cleaning and Painting Airframe	Material ID: **Alternative 1** Operation 3	Material Name: eStrip® GPX Cornstarch Blast Medium, Water, and Tape		
Operations	Mfg. Process	Description	Material Waste	Resources Used	Energy Used	Total
1	Wipe down and wash	Compete wipe down of the airframe is performed to remove any material that would contaminate the blast media. (Severely dirty airframes are washed at this step.)	2	2	0	4
2	Taping and masking	Any crevices are taped off to prevent blast media from contaminating specific areas.	2	1	0	3
3	Blasting	Cornstarch medium is used to blast the paint and other coatings from the substrate.	1	0	1	2
4	Chemical stripping	Chemical stripping is used to remove paint from any remaining areas of the aircraft.	2	1	0	3
5	Thorough wash	Each aircraft gets a thorough wash to remove any blast media after the stripping process is complete.	1	2	0	3
		Total	8	6	1	15

identical, then there cannot be an aggregate total for the PCPC. This problem also prevents the comparison on a plantwide basis of similar processes unless a common system of measures is used in the PCPC scoring. The next application demonstrates a change in the metrics and a very different application of the PCPC concept.

Application 4: Custom Machine Builders

Type of Manufacturer: Manufacturing machine designers and builders

Objective: Improve plant lighting and reduce the utility costs

Custom Machine Builders is located in Ohio. They design and build custom production machines, specifically automated assembly line equipment for manufacturers. The company's production area excluding offices is about 22,500 square feet. The plant houses several pieces of CNC equipment along with an assortment of other machine shop and metalworking equipment. There is also a 4000 square foot assembly area that is surrounded with work benches and parts carts.

The 28-year-old building has a flat roof that provides an unobstructed 14-foot ceiling height. Mounted between the ceiling joists are a variety of lighting fixtures that include incandescent, fluorescent, and sodium vapor lighting fixtures. The sodium vapor lighting gives off a very unnatural yellow hue and is akin to the yellow-orange lighting that is found in parking lots and streetlamps. Maintaining this assortment of lighting, which includes replacing bulbs and tubes, requires about 1.5 hours per week of maintenance work. The fixture spacing is irregular, which creates shadows that make it difficult for the machinists to carry out their work—particularly in the assembly area.

To overcome these problems the plant engineer developed a lighting plan. The plan made use of information provided by the utility company serving the plant. The utility company also referenced standards from the American Society of Heating,

Refrigerating and Air-Conditioning Engineers (ASHRAE). This is an international organization that serves the heating, ventilation, air conditioning, and refrigeration industry. They conduct research and publish standards such as ASHRAE 90.1-2007, which covers lighting and lighting controls. For a low bay (ceiling height of 20 feet or less) manufacturing area the recommended guideline for lighting is 1.3 watts per square foot (see http://www.energycodes.gov/training/pdfs/lighting07.pdf). With proper design this level of power equates to a lighting level of 55–65 foot-candles at a bench top, about the same level of lighting as there would be in a very well lit office.

The lighting specialist at the utility company also encouraged the plant engineer to assess the current lighting system to establish a baseline. The assessment involved taking light level readings at the work surfaces around the plant and then counting the number and type of fixtures. The result showed that the lighting levels varied from a low of 20 to a high of 70 foot-candles. The average level was 45 foot-candles. As a reference, 20 foot-candles is equivalent to the lighting level found in a hotel hallway and 45 foot-candles is about what you would find in the reading area of a library. Finally the total wattage used to power the current lighting fixtures was calculated and determined to be 49.8 kW. This was a conservative figure based on the total of the stated wattage printed on bulbs and tubes that were used in the fixtures.

The engineer entered the baseline power consumption in his adapted version of the PCPC. Next using the guidelines from the plan the company sent out a request for quotation to two electrical supply companies (company A and company B). They received quotes on three types of replacement light fixtures from the suppliers. All three types of lighting fixtures (LFX 1, LFX 2, and LFX 3) that were quoted would meet the foot-candle requirement.

However, installation and purchase costs were just a few of the variables that had to be evaluated. This was due to each fixture's performance in terms of power consumption that affected the watts per square foot, bulb life, and the fixture's expected length of service. As a first step in the process to select a fixture type the

TABLE 7.4.   Cost to Purchase and Install Three Types of Lighting Fixture from Two Suppliers

Fixture Type	Installation Cost	Fixture Quote		Total Cost
		Supplier A	Supplier B	
LFX 1	$12,000	No quote	$42,000	$54,000
LFX 2	$13,800	$43,000	$38,500	$52,300
LFX 3	$11,000	$20,800	$24,000	$31,800

plant engineer put together Table 7.4. This table shows the fixture cost by supplier and the cost of installation.

The shaded area indicates which supplier is being considered as a supplier for a particular fixture. It is understood that once the bid is accepted the price may vary 10% depending on market conditions.

In the past a capital expenditure of this type would generally be determined by the lowest installed cost that meets a performance specification. However, operation costs over the life of the project were also part of the evaluation process. Therefore the next step was putting together a worksheet to determine the annual operating costs for each fixture type. The calculations are predicated on 49 weeks of operation per year at 5 days per week and 10 hours of lighting per workday. The plant engineer included two other entries in this table. The first was a "best practice" entry based on ASHRAE's guidelines and the second was the current lighting system's power requirements.

The ASHRAE guidelines provide a benchmark of 1.3 watts per square foot. For this plant that provides an annual power cost of $4300 per year as noted in Table 7.5. With this information the plant engineer began the preparation of the PCPC (Table 7.6). Since the data being developed is in dollars, the other categories for waste should also be in dollars. Some of this information would come from the maintenance department and some from operations.

Similarly, a PCPC was developed for each of the other two fixture types (LFX 1B and LFX 2B) that were being considered. Those PCPCs are not shown here, but the same format was used.

TABLE 7.5. Power Costs for One Year of Plant Lighting

Fixture	Number of Fixtures Needed	Total Lighting (kW)	kWh/Year at 2450 hours per year	Annual Power Cost at $0.06/kWh
ASHRAE			71,667	$4,300
Plant's Current Lighting		49.8	104,860	$7,321
LFX 1	260	29.0	71,050	$4,263
LFX 2	300	32.5	79,625	$4,778
LFX 3	225	36.0	88,200	$5,292

The unique aspect in these studies is the calculation of the annual waste assessment based on the installation of new lighting fixtures. In the example shown in Table 7.6, the proposed lighting system (LFX 3A) is being compared with the ASHRAE best practice guide. As indicated in bold, the annual waste assessment for this proposal is $992. This is a marker that indicates that more needs to be done. An investment analysis for this fixture, however, will use the $2029 in annual energy savings that is based on the difference between the current lighting system and the proposed system using the LFX 3A fixtures.

Comment The PCPC analysis presented in this application doesn't factor in the costs of maintaining the current system (1.5 hours per week). We also don't know how this project ranks with competing projects for the money the company has available for capital investment. This would have to be included in the development of the justification and analysis of the project. However, the use of the PCPC approach provides a basis for evaluating a project that is unique. The unique aspect in this application is the inclusion of a best practice as a benchmark to assess the project's standing in the minimizing of waste. It serves to challenge the plant engineer and management to find ways to meet or exceed this benchmark.

TABLE 7.6. One of the PCPCs Produced for this Application—Showing One of the Three Fixtures (LFX 3A) Being Considered

Profitable and Compliant Process Chart (Partial)

Project: **Replacing Plant Lighting System**		Work Order # Pending			Material/Supplier: **LFX 3A**	
Item	Operation	Cost	Material Waste	Other Waste	Energy Used	Total
1	Purchase LFX 3A	$20,800				$28,000
2	Install LFX 3A	$11,000	$800*	$50**		$11,850
Total Projected Installation Costs		$31,800	$800	$50	—	**$40,750**

Notes: *Landfill charge for removed fixtures and hazardous waste cost for bulbs, tubes, and non-PCB ballasts.
**Installation materials landfilled other than recycled cardboard from new equipment.

Annual Waste Assessment

Lighting System Comparisons	Current Power	LFX 3A Power	ASHRAE Best Practice	Waste in Dollars
Current system compared with ASHRAE	$7321		$4300	$3021
Current system compared with LFX 3A	$7321	$5292		$2029
Annual Waste Assessment LFX 3A compared with ASHRAE		$5292	$4300	**$992**

Energy waste averted due to this project is indicated in **bold**.

OBSERVATIONS

You have seen how the PCPC can be used as a decision-making tool. The development of worksheets and scoring keys provides a basic and universal methodology for using the technique. As a way to demonstrate its versatility, four other applications in very different situations were presented. The first application used the PCPC as a format for documenting manufacturing operations to establish a baseline. Next the PCPC was used to evaluate two material stocks. Then it was used to gauge the effectiveness of secondary material stocks to reduce the waste generated by a manufacturing process. Finally, we demonstrated how a much-modified PCPC can play a role in gauging waste during the evaluation of a facility improvement.

All too often techniques such as the PCPC become standalone records and worksheets in a project folder. Their utility is not fully realized unless they become part of the ongoing practice of operating and managing a manufacturing facility. Let's stop for a moment and think again about Avionic Systems, Incorporated, the first of the four applications presented. In that application the photographs showing setups for testing and the assembly processes were inserted in the PCPC to illustrate each operation. This was useful in showing what was taking place so the people involved in the review could make suggestions based on a clearer understanding of the operation. As mentioned in the comments for that application, the photos could even be videos if the PCPC was on a tablet or laptop computer projecting the PCPC on a screen.

Constructing the Virtual PCPC

This suggests that one way to improve the amount of information provided in the PCPC is to create what we call the *virtual PCPC*. This could be done by creating the PCPC using Microsoft Word or Excel and inserting video clips into the document. The key here is for an in-house technician to record the videos and either insert short clips in a single computer or place them on the company's

Intranet. Then the video is available to describe what occurs at the decision point under review. The length of each clip should be about one minute. With this approach members of the review team can actually watch and study the operation taking place from their own desktop. This also means that they do not have to physically be meeting at the same time and in the same place as the rest of the review team to do the preliminary analysis. We call this the A^3 (Anytime Anywhere Approach) for continuous improvement.

To capitalize on the advantages of the virtual PCPC there has to be a means for the responsible groups to access the videos and have a way to communicate and post comments. As noted the video clips can be made available on a server. In many cases a company can utilize its own server for this purpose. Of course, a major advantage for using the private web server is to provide security of access to the information that is online. If a company Intranet server is not available and the process is not confidential, then YouTube, the video-sharing website, can be used free of charge (see www.youtube.com). The second requirement deals with the means to view the video. One approach is to install the Adobe Flash Player plug-in for your web browser or rely on the computer's media player. Methods for posting comments can be as simple as email or chat tools. One tool that we enjoy using is the open source tool for creating a website or blog and task management system called WordPress (see http://wordpress.org/). While the software is free for use, companies would purchase the secure version to ensure privacy and confidentiality of information.

The tools to be selected should be operated using a secure server to preserve access to confidential information. One tool that has proven its usefulness in creating a secure virtual chat network is called mIRC (Mirrored Internet Relay Chat). This can be password protected and run from the company's Intranet server. This tool has been used by the authors for many years in distance education courses, teaching, interacting, and solving problems with diverse groups.

CONCLUSION

There are people all over the globe working on waste reduction and the conservation of resources. Many of these individuals work in the field of manufacturing. For them profitable manufacturing operations are essential to advance technologies and practices that are necessary to protect the environment and conserve resources. If you are involved in manufacturing, then you share some of this responsibility.

Early in this book we posited that waste is a term that includes all those things that do not add value to a product. Therefore eliminating waste reduces costs and improves a company's potential for profitability without diminishing their product's value. Emissions, solid waste going to a landfill, and energy that is expended beyond the minimum needed are other forms of waste that add nothing to the value of the product. Therefore waste in all these forms thwarts a company's efforts to be profitable as well as being environmentally compliant.

One more thing ought to be mentioned. We know that every company wants to reduce waste to become profitable. Getting started is the difficult part, but this is where the fun really begins. You can use any and all the tools and techniques we have mentioned and even develop your own as well. Just make sure that you don't miss the decision points that provide the opportunities to design out waste. Also develop your own metrics and scoring keys for the worksheets that you will need for your company. Establish a baseline for continuous improvement. Look for world-class benchmarks to establish your goals.

If you haven't adopted continuous improvement as a basis for operations, then all of this may seem almost overwhelming. This is to be expected because there are many things to consider and a lot of data to be collected. Human nature is to try to do everything at once. The way to break free from this thinking and reduce the chance of a false start or procrastination is to stop for a minute to realize that complex undertakings can often be improved by systems thinking. Remember the words of Albert Einstein when he said "make everything as simple as possible, but not simpler." Systems thinking requires concentrating on parts of the system and then dealing with them one step at a time. It's a linear process that does not have to be difficult. The beauty of it all is that the process itself helps us to gain momentum. Once again, recall from the case study chapter that the Kaizen approach is based on a series of simple steps done continuously. One task well done provides reinforcement and energy to tackle the next task with more vigor. Once you finish you are at the beginning of continuous improvement. Keep your good work going. Enjoy this exciting adventure!

SELECTED BIBLIOGRAPHY

Google. Chrome browser. Available at https://www.google.com/chrome (accessed January 18, 2012).

U.S. DoE. (2008). ANSI/ASHRAE/IESNA 90.1-2007: an overview of the lighting and power requirements. Washington, D.C.: U.S. Department of Energy, Office of Energy Efficiency and Renewable Energy. Available at http://www.energycodes.gov/training/pdfs/lighting07.pdf (accessed January 18, 2012). [Overview of ASHRAE 90.1-2007.]

YouTube. Video hosting. Available at www.youtube.com (accessed January 18, 2012).

GLOSSARY

5S Process   A Japanese process developed to describe how to organize a work space to improve productivity by identifying and storing items in a specific location and then preserving the new order. The five steps involved are *Seiri* (Sort), *Seiton* (Straighten), *Seiso* (Shine), *Seiketsu* (Standardize), and *Shitsuke* (Sustain).

A³   The "Anytime Anywhere Approach" for continuous improvement.

Basic Process Classifications (BPCs)   There are five basic process classifications: forming, separating, joining, conditioning, and finishing.

Building Life Cycle Analysis   Used to assess the impact a building will have on the environment and the business.

Can of Worms   A complex troublesome situation with a mess of entanglements that cause extensive subsequent problems when a decision or action is taken.

Improving Profitability Through Green Manufacturing: Creating a Profitable and Environmentally Compliant Manufacturing Facility, First Edition.
David R. Hillis and J. Barry DuVall.
© 2012 John Wiley & Sons, Inc. Published 2012 by John Wiley & Sons, Inc.

Case Study   An explanatory analysis identifying what an organization or facility has done to achieve an objective.

Conditioning Processes   Processes that produce a change in the mechanical properties or molecular structure of a material.

Continuous Flow Manufacturing   Aimed at minimizing work-in-process and reducing lead times, thus making the manufacturing system more reactive to the customer.

Cooling Degree Day (CDD)   Used as a measure of energy consumed for cooling. A day with an average temperature 1 degree above 65°F (18.33°C) is one cooling degree day. For example, a day with an average temperature of 80°F is equal to 15 cooling degree days.

Decision Points   Opportunities for waste reduction or resource minimization.

Design for Assembly   An approach that outlines how products should be designed for either manual or automated assembly. It also explains how to design a product to eliminate errors.

Design of Experiments (DOE)   Used to determine the significant variables that affect a specific outcome of a process or operation. Once the significant effects are identified it enables the operators to control and reduce the variability of the process. This approach to experimentation allows for the investigation of several variables at once. It also highlights the presence of interaction between variables, which is often one of the most confounding problems in manufacturing.

DMAIC   Associated with DOE, this term stands for "Define, Measure, Analyze, Improve, and Control," the names of the five steps that one follows in its application. It is one of several methods used to design and carry out process improvement.

Durable Good   A hard good that lasts for three or more years. It is a product that is not completely consumed through use and continues to provide value.

Early Manufacturing Involvement   A technique that involves personnel from manufacturing in the design of a new product.

Easily Adaptable Building (EAB) A building configuration that enables a manufacturing facility to be changed over to produce a different product easily with minimal downtime.

Energy Performance Indicators (EnPIs) Quantitative measures that are used to determine the effectiveness of an energy management plan.

Finishing Processes Processes that clean the product and/or apply a coating that can serve to protect it or improve its appearance.

Five Whys Method A process that involves asking why until the root cause of a problem is determined. Shigeo Shingo, an industrial engineer working for Toyota, popularized this technique.

Flexible Workforce A workforce that is able to respond just-in-time (JIT) to changes (downturns and upturns) in production levels and do it without generating waste. The JIT workforce retains an organization's knowledge, skills, best practice, and culture that are essential to produce products without generating waste.

Forming Processes Processes that are used to shape, stretch, twist, bend, and/or compress materials to a desired form.

Global Industry Classification Standard A major system for equities developed jointly by Morgan Stanley Capital International (MSCI) and Standard & Poor's.

Group 2 A Stage 2 manufacturer that builds products that others design.

Heating Degree Day (HDD) Used as a measure of energy consumed for heating. A day with an average temperature 1 degree below 65°F (18.33°C). As an example, a day with an average temperature of 60°F will be equal to 5 heating degree days.

Horizontally Integrated Manufacturer A manufacturer that is involved in just one or two steps in the manufacturing sequence. Frequently such companies make more than one product.

Industrial Production A term that is used by the Federal Reserve Board (FRB) to refer to the total output of U.S. factories and mines and is a key economic indicator for the U.S. economy.

Industry Classification Benchmark A system that is popular in the fields of finance and market research. It was developed by Dow Jones and the FTSE Group (which is often known by its nickname "the Footsie").

Interval Scale A scale used for measurable data, such as temperature in degrees Celsius.

JIT Workforce A staffing method a manufacturing company can use to create a flexible workforce. See the definition for a flexible workforce.

Joining Processes Processes that are used to fasten or fabricate industrial stock, parts, subassemblies, components, or partially finished workpieces together. There are three types: adhesion, cohesion, and mechanical joining.

Just-in-Time (JIT) Describes a production system in which a manufacturing company does not maintain a large inventory of materials. JIT was intended to foster the creation of more-responsive operations across all stages of manufacturing.

Kaizen Japanese word meaning "improvement" or "change for the better." It has become more widely known as the process of *continuous improvement*. In practice it has been used as the basis for reducing costs and improving service in virtually every form of business activity.

Lean Manufacturing A comprehensive approach to waste reduction in manufacturing.

Life Cycle Analysis A technique to assess the environmental impact of a product or a facility through all stages of its life.

Life Cycle Cost Analysis A method used to assess the costs of a facility or product through all stages of its life, from its design until it is no longer in use and is torn down. It also includes the cost of recycling the building or product to either create a new stock or, if it is unable to be recycled, then its proper disposal.

Load Values Numerical values that appear in four categories on each of the worksheets used to develop the PCPC. Load values represent the potential for creating waste: the higher the value, the greater the potential.

Major Manufacturing Enterprises (MMEs) The companies or individual firms that manufacture products from wood, metal, ceramic, plastic and/or composite materials.

Major Material Families (MMFs) Woods, metals, ceramics, plastics (including elastomerics), and composites.

Major Product Groups (MPGs) Groups based on types of products, used to classify manufacturing establishments. According to NAICS, there are 21 major product groups for manufacturing, based on types of products; four of these are concerned with hard good manufactured products.

Manufacturing Facility Sustainability Refers to a building's ablity to hold up, retain its utility, and be adaptable to changes in manufacturing technology.

Manufacturing Sequence The sequence or steps involved in the creation of a product.

Manufacturing Specialist A manufacturer whose production operations focus on one major material family and process: for example, a plant with one primary process such as a range of injection molding machines that produce molded plastic parts from a range of primary stocks, thermoset or thermoplastic materials.

Material Drops Materials that are removed through processing and are not part of the final product.

Material Selection Worksheet A worksheet used to collect data on selected material stocks to identify which candidate is best for use in terms of product performance, profit potential, and minimum impact on the environment.

Maximum Achievable Control Technology (MACT) Technology that provides a means to use materials that have been listed as toxic or polluting.

Nominal Scale A categorical scale that identifies things such as the names of the 50 states of the United States.

Nondurable Goods Soft goods that are immediately consumed, wear out, or have a life cycle of less than three years.

North American Industry Classification System (NAICS) An industry classification system adopted in 1997 that was developed by the Office of Management and Budget (OMB) in collaboration with Canada's U.S. Economic Classification Policy Committee (ECPC), Statistics Canada, and Mexico's Instituto Nacional de Estadística y Georgrafía. NAICS (pronounced "Nakes") is used by statistical agencies to codify business and industrial establishments when collecting economic data.

North American Product Classification System (NAPCS) A classification system that is currently being developed to integrate all of the industries defined in NAICS in terms of their products. The emphasis will be on goods produced and services provided.

On-Call Workforce (OCW) Refers to a flexible or JIT workforce.

Operator Self-Control A term that defines the fundamental work objectives for every individual in the business of manufacturing. This is the basic building block for management. Joseph M. Juran and Frank Gryna developed this concept when the United States was struggling to overcome the problems plaguing their attempts to place satellites in orbit.

Ordinal Scale A ranking scale, such as 1st, 2nd, or 3rd place.

Pareto Principle Developed by Vilfredo *Pareto*, an Italian economist who in the early part of the twentieth century observed that there always seemed to be a disproportionate allocation of the ownership of land. Pareto noted that 80% of the land was owned by only 20% of the population. This has become known as the 80/20 distribution.

Pick the Low Hanging Fruit An approach emphasizing doing the easiest tasks first.

Plan, Do, Check, and Act (PDCA) A management process that has been in wide use and is now part of the approach to energy reduction used by the International Organization for Standardization (ISO).

Plant Location Analysis   An analysis of the costs and supporting infrastructure for selecting a plant site. Some of the factors in the analysis are ease of transportation, cost of living, and proximity to suppliers.

Poka-Yoke   A Japanese term meaning "fail-safe" or "mistake-proofing." Application of this concept to processes, machines, and work sequences means they would be designed to eliminate the possibility of making a mistake.

Primary Manufacturing Industry   Stage 1 manufacturers (extraction industries) that produce the first form of a stock from raw materials.

Process Selection Worksheet   A worksheet used to collect data on selected processes to identify which of several candidates is most effective in minimizing waste, resources, and its impact on the environment.

Product Life Cycle   Refers to the closed loop of the manufacturing sequence that begins with the raw material and ends with recycling after the product has been discarded.

Product Life Cycle Management   Refers to the strategies a business uses to manage the life of a product in the marketplace.

Product Lifecycle Management   The management of the information acquired over a product's life so that the manufacturer understands how the product is designed, built, and serviced.

Profitable and Compliant Process Chart (PCPC)   A methodology and a decision-making tool that can be used by all the functional groups in an organization. For new products, departments such as marketing, design engineering, and manufacturing engineering can use this methodology as a basis for collaboration to minimize waste that might be inherent in a product design. It can also be used by the groups involved in manufacturing existing products to reduce or minimize waste.

Ratio Scale   A scale is used to measure length, mass, energy, and so forth. All statistical measures can be used for data that is represented on the ratio scale.

Raw Material   The basic component used to create a manufacturing stock.

Recycled   Refers to a product that has been discarded and will be processed to become a manufacturing stock.

Responsible Care Codes of Practice   The Responsible Care Codes of Practice is a chemical industry initiative to continuously improve a facility's operations in terms of health, safety, and compliance with environmental standards.

Secondary Manufacturing Industry   Stage 2 industries that use the stock(s) from primary manufacturing industries to produce hard good consumer products.

Secondary Material   A material that is needed in the production of a product but does not become part of the product.

Separating Processes   Processes that are used to remove materials or volume with or without the creation of a chip.

Single-Minute Exchange of Dies (SMED)   A term that describes a rapid changeover of equipment. This concept implies that retooling or changing over a piece of equipment happens so quickly that there is only a minute's worth of downtime.

Situational Awareness   Refers to being aware of factors that can affect a product or manufacturing facility, such as impending changes in technologies, as well as relevant regulatory obligations.

Six Sigma Approach   An approach to manufacturing that was developed as way to eliminate defects and the problems they cause. It is a way of thinking using a system of skills and methods to create a manufacturing operation that is very close to being able to produce defect-free products. The term "six sigma" refers to a manufacturing process that is in control and subject only to normal variability. The degree of process control is such that the process mean is separated by six standard deviations or more from each specification limit.

Smart Meter   A device that can provide data on where, when, and how a resource is being consumed.

Sources and Uses   An analysis to identify the source and amount of a resource or stock that is being purchased and determining where and how it is used.

Stage 1 Manufacturing   Refers to companies that are engaged in primary manufacturing.

Stage 2 Manufacturing   Refers to companies that use manufactured stock(s) to produce finished products and components.

Stage 3 Manufacturing   Refers to companies that deal with distribution, service, reclamation, and disposal of products.

Standard Industrial Classification Index (SIC)   The first standard industrial classification for the United States, adopted in 1939. The first List of Industries for manufacturing in the United States was published in 1938. This list was consolidated with the 1939 List of Industries for nonmanufacturing industries and became the first standard industrial classification for the United States.

Stevens Power Law   Developed by Stanley Stevens. The law states that there are four types of scales that can be used to define how things can be measured or arranged: nominal, ordinal, interval, and ratio.

Stock(s)   The raw materials supplied to manufacturers in forms such as sheets, bars, wire, rods, pellets, powders, and liquids that are used to make products.

Supplier Selection Worksheet   A worksheet used to collect data on suggested suppliers to identify the best of several candidates in terms of their performance as a quality and environmentally conscious supplier.

Sustainable Practice   Refers to activities that a manufacturing organization uses to be profitable and environmentally compliant.

Tertiary Manufacturing Industry   Refers to industries that concentrate on product distribution, sales and service, and disposal of worn-out products.

Three-Sector Hypothesis   A hypothesis developed by Jean Fourastié that industrial activity occurs in three stages or sectors:

stage 1, primary manufacturing; stage 2, secondary manufacturing; and stage 3, distribution and service of manufactured products.

Total Quality Management An integrated management effort to continuously improve the quality of its products.

Value-Added Potential A measure of the value that can be added at each step in the manufacturing sequence.

Vertically Integrated Manufacturer A manufacturer that is involved in three of more steps in the manufacturing sequence.

Virtual PCPC The profitable and compliant process chart as a digital file that can access an engineering database and other manufacturing performance measures. The virtual PCPC can contain video clips of manufacturing processes and operations that can be viewed by team members using the Anytime Anywhere Approach.

Work-in-Process (WIP) Products or components in the process of being manufactured. This includes idle product that is being stored or waiting to be worked on.

World Class Manufacturing Used to describe manufacturing systems. This term embodies a range of activities. In the 1980s the term was used to define manufacturing systems that pared away any activity that did not add value to the product being produced. This practice evolved into the elimination of wasted effort, time, and material that did not create value for the customer. Product that was not being worked on or on its way to the next operation that would add value to it was considered to be a waste of production resources.

INDEX

Improving Profitability Through Green Manufacturing: Creating a Profitable and Environmentally Compliant Manufacturing Facility, First Edition.
David R. Hillis and J. Barry DuVall.
© 2012 John Wiley & Sons, Inc. Published 2012 by John Wiley & Sons, Inc.